The Tech Prep

Associate Degree

Challenge

A Report of the Tech Prep Roundtable

A Project of the American Association of Community Colleges with Funding from The Boeing Company

Tech Prep Roundtable
James McKenney, Project Director

The Tech Prep Associate Degree Challenge
Lisa Falcone and Robert Mundhenk, editors

American Association of Community Colleges
National Center for Higher Education
One Dupont Circle NW, Suite 410
Washington, DC 20036
(202) 728-0200

Additional copies of this publication may be purchased from:
AACC Publications
P.O. Box 311
Annapolis Junction, MD 20701
(301) 490-8116

TABLE OF CONTENTS

Part I

FOREWORD

T HE CONTRIBUTORS TO THIS PUBLICATION ARE ACKNOWLEDGED TO be national leaders in occupational education. They represent universities, community colleges, schools, industry, labor, and the federal government. Their willingness to dedicate their expertise and time to the discussion and advancement of Tech Prep Associate Degree programs at community colleges demonstrates the importance they place on this critical educational reform.

The Boeing Company's involvement in Tech Prep Associate Degree programs stems from the nation's need to recognize and address the skills and employability of the nearly 80 percent of high school graduates who likely will not complete a baccalaureate degree. When these students leave high school, they generally are not prepared for entry into high-skilled jobs. Until recently educational reform efforts paid little attention to this segment of the school population, even though the national demand for technically skilled workers is on the increase.

The essence of Tech Prep is to provide young people with the educational option to combine learning with doing. Most of us learn best within a contextual framework, but, unfortunately, a contextual learning approach typically is not emphasized in many schools. Theoretical approaches, which are designed to prepare students for baccalaureate degree programs, are predominant in most U.S. high schools, underscoring the importance American society places on the four-year degree. The implicit message, however, is that anything less than a baccalaureate degree represents mediocrity or even failure. The prevalence of this misconception convinced The Boeing Company that Tech Prep's focus on contextual learning and applied academics would contribute to the skill needs of our industry, while enhancing the well-being of our communities.

It is my experience that when business representatives learn about Tech Prep's attributes they endorse it wholeheartedly. When educators understand the value of providing school-based and work-based learning opportunities to students through articulated high school and college programs, they also support the merits of TPAD programs. One highly respected school superintendent from eastern Washington said that he regards Tech Prep as the most revolutionary change in pre-baccalaureate curricula in fifty years. Similar comments have been made by other educators, as well as a wide array of business executives.

Tech Prep is education reform in the truest sense, and it underscores what tends to be missing in other reform movements—a clear and distinct focus on teaching and learning. It is also a powerful model of the essence and spirit of the School-to-Work Opportunities Act. The challenge of Tech Prep now is to institutionalize the model throughout the nation. The publication you are about to read is an important step toward that vision.

Carver C. Gayton
Corporate Director
College and University Relations
The Boeing Company

Part II

PREFACE

L IKE OTHER EDUCATIONAL REFORMS THAT HAVE GONE BEFORE IT, Tech Prep is at a critical crossroads in its development. The underlying concept of Tech Prep is solid, promising focused educational and training opportunities for any student who desires to learn and achieve. However, challenges with the implementation process persist, particularly at the community college level. Developing solutions to overcome these challenges must be accomplished if the lofty vision offered by Tech Prep is to be fully achieved.

In the fall of 1993, the American Association of Community Colleges (AACC) convened some of the nation's foremost Tech Prep authorities for the Tech Prep Roundtable, a two-day meeting in Washington, DC. The purpose of the meeting was to discuss the direction Tech Prep has taken since its enactment through the Perkins reauthorization in 1990 and the emerging issues related to successful implementation of Tech Prep Associate Degree (TPAD) programs at community colleges. Roundtable participants represented the U.S. departments of Education and Labor, the National Center for Research in Vocational Education, the Center for Occupational Research and Development, community colleges, secondary schools, and the business community. Prior to the roundtable, six of the participants were commissioned to write white papers that examined the critical implementation and policy issues affecting the success or failure of TPAD programs. Presented at the meeting by their authors, these papers served as a catalyst for discussion and debate among the group.

Participants agreed that Tech Prep is one of the most promising education and training reforms in recent American history, with the

3

potential to make a dramatic difference in the lives of numerous students. By revolutionizing traditional curricula and pedagogy, the Tech Prep Associate Degree offers students a clear, practical connection that ties their educational experience to a career pathway. It is viewed as essential to both secondary and postsecondary education reform, and to the nation's economic development strategy. The formidable task now is to ensure that TPAD programs are properly developed and implemented across the country.

This publication, a product of the Tech Prep Roundtable, is designed to assist community colleges in their efforts to implement Tech Prep Associate Degree programs. It contains a broad overview of Tech Prep, a series of recommendations for implementing TPAD programs at community colleges, and the six white papers presented at the roundtable. Although TPAD programs are typically developed in consortia, this publication, particularly its recommendations, is aimed at a community college audience and emphasizes the community college role in the process.

Special thanks are extended to The Boeing Company for funding both the Tech Prep Roundtable and this publication, to Digital Equipment Corporation for providing financial support for the roundtable, to the roundtable participants for dedicating both their time and expertise to the process, to James McKenney for coordinating these Tech Prep activities, and to Robert Mundhenk of Northampton County Area Community College and Lisa Falcone for providing editorial assistance for this publication.

David Pierce
President
American Association of Community Colleges

Part III

Tech Prep Roundtable Participants

Jacquelyn M. Belcher
Minneapolis Community College, Minnesota

Ron Castaldi
U.S. Department of Education

Robert Darden
Digital Equipment Corporation

Don C. Garrison
Tri-County Technical College, South Carolina

Carver C. Gayton*
The Boeing Company

Carolyn Graham
Charles County Board of Education, Maryland

Gerald C. Hayward*
National Center for Research in Vocational Education

Dan Hull
Center for Occupational Research and Development

Richard Kazis*
Jobs for the Future

Wm. Carroll Marsalis
Tennessee Valley Authority

Laurel McFarland
The Brookings Institute

Michael C. Morrison
North Iowa Area Community College

Dale Parnell*
Oregon State University

Jack H. Rapport
U.S. Department of Labor

Raymond Taylor
Association of Community College Trustees

Nellie Thorogood
North Harris Montgomery Community College District, Texas

Diana Walter*
Partnership for Academic and Career Education

Winifred I. Warnat*
U.S. Department of Education

Jerome Wartgow
Colorado Community College and Occupational Education System

AACC Representatives

David Pierce
President and Chairman

James McKenney
Director of Economic Development and Project Director

*indicates author of a white paper

Part IV

IMPLEMENTING TECH PREP ASSOCIATE DEGREE PROGRAMS AT COMMUNITY COLLEGES: SUMMARY AND RECOMMENDATIONS

I N THE EARLY '80S, A NATIONAL STUDY REVEALED THAT INCREASING numbers of American youth were leaving high schools unprepared for either college or work (Gardner et al., 1983). At the time, public high schools typically offered students three paths of study: college preparatory, vocational education, and general education. Of these pathways, a major dichotomy existed between the education and satisfaction level of general education and college prep students. General education students, who often selected personal or hobby-type classes in place of traditional academic subjects, were less satisfied with their academic experience than students in college prep or vocational education. Most alarming, 63.5 percent of all high school dropouts were last enrolled in a general education program before leaving school (National Center for Education Statistics, 1984).

Despite these findings, people involved in education reform initiatives were primarily concerned with strengthening academic education, which was viewed by many as the most promising way to prepare students for college and careers. Increasing standards, tightening graduation requirements, and initiating standardized tests were some approaches being considered to reform schools at the time. These tactics, however, did not attempt to change vocational education or the unstructured general education curriculum that was prevalent in public high schools.

In 1985, Dale Parnell, then-president of the American Association of Community and Junior Colleges, published *The Neglected Majority*. Through this book and Parnell's efforts, national attention was redirected to the majority of the nation's high school students and to the ideas of connecting education to career pathways and integrated learning.

Parnell contended that special needs and talented students deplete most public school resources, including teachers' time and energy. Those in the middle quartiles typically receive minimal attention and often receive an unfocused education that does not prepare them for college or work. As a result, many of these students become lost within the system (Parnell, 1985).

Parnell designed the TPAD concept for this unserved majority of high school students, particularly the three out of four students not likely ever to complete a four-year college degree. Through an integrated applied academic and vocational curriculum approach, a Tech Prep education increases students' motivation to learn, encourages academic success, and prepares them to work in the high-performance workplace of tomorrow. It also addresses society's need for more technically sophisticated workers in a technology- and information-driven economy by providing students with a high-quality educational choice.

This educational reform was designed to include a restructuring of both high school and community college curricula. It assumes curriculum connections between instructional activity and the skill requirements of the work world. It emphasizes contextual learning and continuity in learning by allowing students to choose career pathways predicated on a connected high school and community college course of study. TPAD also encourages participation in work-based learning experiences to enhance the school-based curricula.

Under Tech Prep, technical curricula in high school develop the broad skills for a particular career upon graduation, or for advanced matriculation within a community college technical career program. Community college curricula build on these broad skills to provide advanced technical skills appropriate to employment requirements. Representatives from high schools, community colleges, and business and industry are expected to collaborate on curriculum, as well as program and evaluation design.

In 1990, Tech Prep was codified into federal law through the reauthorization of the Carl D. Perkins Vocational and Applied Technology Education Act. This legislation funded planning and demonstration grants to local consortia for the purpose of developing and implementing Tech Prep programs. The legislation was written broadly enough to accommodate the many differences among the states, but it defines Tech Prep fairly specifically:

> Any such program shall...consist of the 2 years of secondary school preceding graduation and 2 years of higher education, or an apprenticeship program of at least 2 years following secondary instruction, with a common core of required proficiency in mathematics, science, communications, and technologies designed to lead to an associate degree or certificate in a specific career field (U.S. Congress, 1990, pp. 52–53).

Since its adoption, nearly 1,000 consortia and about 5,700 high schools across the country have embraced the Tech Prep model, most with great success. Many community colleges are making Tech Prep central to their institutional missions, viewing it as a four-year program with the associate degree being the capstone that assures standards of excellence are met.

The most recent new influences on the Tech Prep education movement have been the distribution of planning funds by the federal government to states for designing school-to-work transition systems, and the passage of the School-to-Work Opportunities (STWO) Act. While some fear that this recent Clinton administration initiative will undermine TPAD programs, the stated intent of STWO is to strengthen TPAD programs and other initiatives that emphasize the connection between education and careers. STWO fosters a smooth transition between school and work and adds a work-based learning component to the educational experience. Thus, the STWO Act should complement, not work against or in competition with, any TPAD program. With congressional and administrative endorsements and support for TPAD, its future seems secure, unless those designing and implementing TPAD programs fail to live up to its promise. In the spirit of helping community colleges succeed in TPAD implementation, the following recommendations are offered.

REFORMING CURRICULA AND TEACHING

Integrating Applied Academic and Vocational Education
A crucial component of successful TPAD programs is the integration of applied academic and vocational education, an important,

underlying principle of the initial design and legislation. To integrate learning, educators must move from traditional classroom teaching and learning methods to contextual learning, which recognizes that students learn more easily when content is presented in a way that is consistent with the context of its application (Hull, 1993). Students exposed to real-world applications in their learning environments are more likely to understand and retain knowledge because they can see the usefulness of those ideas.

In addition, integrating applied academic and vocational education requires a reformed curriculum and a systemic change in the way teaching is conducted. Curricula of academic courses must connect theories and concepts with life experiences, and occupational curricula should be designed to show the importance of general ideas and knowledge beyond the technical skills of occupations.

Changes in teaching and curricula are already underway at the secondary level, but they are moving at a slower pace at the community college level. If change at the postsecondary level lags too far behind, students who enter a community college after being enrolled in a TPAD program in high school may be frustrated by their new learning environment and lose motivation. Community colleges need to meet this challenge.

Recommendations
1. Community colleges not already doing so should initiate curricular and pedagogical reforms appropriate to TPAD programs.
2. Curricula for TPAD programs should integrate applied academic and technical education wherever possible.
3. Instructors of academic courses should include ample experiences that connect learning to the learner's communities, including work, politics, social issues, and institutions.
4. Curricula for occupational courses should be predicated on the assumption that occupational instructors will teach beyond technical skills and will support the importance of academic education in preparing students for an occupation.

Developing Core Curriculum, Sequencing Courses, and Coordinating Curriculum Planning
Graduates of TPAD programs should leave college with higher-order skills than non-TPAD students or students with only a high school education. In order for workers to obtain these skills, careful

attention must be paid to the design of curricula and sequence of courses.

The TPAD curriculum cannot simply repackage traditional vocational education programs; it should engage students in active learning, closely linked to real-life applications and experiences through applied learning approaches. The curriculum needs to provide contextual learning within a curriculum core that includes both academic and technical education courses. The academic education courses should be common to all curricula and include knowledge necessary to live competently as a citizen, life-long learner, consumer, and producer. Courses in the technical core should include technical skills that are common to a given career pathway, like computer literacy and the role of technology in society.

To achieve desired outcomes efficiently and to avoid unnecessary overlap and duplication, course sequencing between articulated high school and community college programs is essential. This means that technical courses in a community college TPAD curriculum should be built in logical sequence as an extension of the articulated high school TPAD curriculum. Course sequencing is the most important means of raising expectations of students and helping them develop higher-order technical skills. Sequencing of courses requires collaboration of high school and community college instructors, assuring that the connected curriculum structure is program-to-program, rather than course-to-course. Finally, to ascertain if curricula are designed appropriately to prepare students for the workplace, TPAD partnerships should include representatives from the business community, particularly in the planning and design of curriculum.

Recommendations

5. A core curriculum that includes both academic and technical education courses should be developed in collaboration with high school and community college instructors and curriculum specialists. The core curriculum should be common to all occupational programs, although the technical courses might vary between career pathways.
6. Courses necessary to complete a technical specialization should be sequenced in order to ensure the attainment of higher-order skills.
7. Community college TPAD curricula should be developed in collaboration with both high school instructors and representatives from the business community. The curricula should be sequenced and should avoid needless overlap.

Designing Bridge Programs

The majority of community college students are adult students who typically do not enroll in college directly from high school and more than likely did not participate in a high school TPAD program. When they arrive at college, they are often unprepared for the challenging curricula in most TPAD programs and need special assistance. "Bridge" programs are necessary for these students as well as traditional-age community college students who did not take Tech Prep courses in high school. Tech Prep Associate Degree programs are most often based on an articulated curriculum, consisting of two years at both the high school and community college levels. Because of this, it is critical that bridge programs ensure that students who did not have the high school component receive appropriate courses and preparatory services to prepare them for the more advanced levels of a TPAD program.

Community colleges should work with feeder high schools to develop bridge curricula patterned on the high school curricula, with appropriate academic and technical foundations for the community college program. Colleges should assess these students' knowledge base and relevant skill level at the time of enrollment.

Recommendations

8. A bridge curriculum should be designed to provide for adult entry into technical programs. This bridge should be based on the academic, technical, pedagogical, and counseling components of the high school program that have been articulated with the community college technical program.
9. Entering adult students should be assessed to determine their knowledge and skill level relative to the technical program they select for study.

Improving Instruction and Student Services Through Professional Development

Integrating applied academic and vocational education through TPAD programs requires instructors to reach beyond the confines of traditional classroom teaching models and adopt new pedagogical techniques.

At the high school level, great strides in changing and updating teaching methods have been made by both vocational and academic-course instructors to create a more contextual learning environment for students. Desired learning outcomes are being achieved when teachers incorporate new contextual teaching techniques and experiences, including work-based learning opportunities. As a result, early assessments of TPAD success suggest that high school students are embracing the Tech Prep choice and are becoming excited about learning.

Contextual learning in TPAD programs should not be the exception, but rather the rule—both in high schools and community colleges. As students complete high school Tech Prep programs that have been based on applied academics and contextual learning, they expect and anticipate that a similar learning environment will exist in community college. Unfortunately, indicators show that the pace of change to applied academics and contextual learning in community colleges needs to increase substantially to prepare for, and accommodate properly, students who successfully complete TPAD programs in high school.

To facilitate this change, institutions should provide professional development opportunities for instructors, as well as for career counselors and other TPAD program staff who play an integral role in the education of students. Ongoing professional training for instructors should provide them with the skills to incorporate contextual and participatory learning in their teaching methods. Professional development for all other TPAD staff should ensure that they are knowledgeable about the changes in curricula as well as workforce demographics and needs, so that they are better able to assist students with their academic and career decisions. Professional development programs and opportunities for instructors and staff will enhance the quality and effectiveness of TPAD programs.

Recommendations

10. Instructors in both academic and occupational programs should be provided with training that will enable them to utilize contextual learning techniques to maximize student understanding.
11. Continuous, comprehensive staff development should be provided to community college career counselors, as well as other TPAD program staff.

EVALUATING TPAD PROGRAMS

Defining Learning Outcomes

An important part of TPAD curriculum design is the sequencing of courses and the definition of learning outcomes. It is essential that student outcomes be identified early in the development process and be measured against clearly stated goals. To assure that the goals and outcomes are relevant, consultation with representatives of the business community is necessary.

It is clear that technical skill outcomes are important in measuring work-based performance. However, the outcomes that address basic skills, team skills, and other essential skills like communications, human relations, and work attitudes are too frequently ignored. Because TPAD programs should produce graduates capable of dealing with a wide range of job challenges, not a narrow grouping of technical skills, a holistic approach to determining outcomes should be a primary goal of TPAD programs.

Recommendation

12. Occupational programs should be designed to achieve defined outcomes that address required technical, work, and other educational skills. As appropriate, consideration should be given to recommendations and/or requirements outlined in *What Work Requires of Schools: A SCANS Report for America 2000*, Goals 2000: Educate America Act, and other applicable reports or legislation.

Developing Performance Indicators

As required in the Perkins Act of 1990, states are responsible for designing their own system of standards to evaluate programs and ensure accountability. Each TPAD program, because it is accountable for student outcomes and program effectiveness, must design a process for evaluation. This process should be developed in partnership with high schools, business representatives, and local and state government officials.

Evaluations based on understandable "performance indicators" should be conducted on a regular basis. Because these performance indicators provide the foundation for monitoring the success of a TPAD program, they should be measured against clearly defined goals that reflect the new standards for connecting education with the competencies needed by employees in the workplace. They should measure change in both the student and program, both in the areas of academic achievement and work-based competencies. Some examples of performance indicators are technical skills, academic learning, connection with work, placement success, employer satisfaction, job satisfaction, retention success, and enrollments in postsecondary programs. Evaluations should be viewed as an essential component of all TPAD programs.

Recommendation

13. Indicators of success should be developed for TPAD programs and a policy of periodic assessment should be adopted. Corrective action should be taken whenever phases of the TPAD program fail to achieve the levels of performance established by the indicators.

CREATING SCHOOL-TO-WORK LEARNING OPPORTUNITIES

Building Education and Employer Partnerships

The success or failure of a Tech Prep Associate Degree program may very well depend on the strength of the connection between the education and business communities. In order to blend academic and vocational curricula with the world of work, full partnerships between education and employers must be established and maintained in all aspects of Tech Prep, including curriculum design, work-based learning opportunities, and program evaluation.

The Perkins Act of 1990 requires that a consortia be established for the purpose of developing TPAD programs. When schools design or modify TPAD programs, they should include representatives from the business community. For students to gain the necessary skills for a changing workplace, business and industry representatives must participate on planning committees and be full partners in the development, implementation, and evaluation of curricula and programs.

Recommendation

14. Business and industry representatives should be included as full partners with high school and college personnel in assessing program need, developing program curriculum, and in assessing program effectiveness.

Incorporating Work-Based Learning and Other School-to-Work Initiatives into TPAD Programs

As states prepare their school-to-work plans, it is imperative that TPAD program administrators and instructors incorporate work-based learning opportunities, whenever possible. Apprenticeships, co-op learning, on-the-job training, and career academies are all forms of work-based learning. Perhaps the best example of where community colleges have successfully incorporated a work-based component is in associate degree nursing programs. Because it is critical to student success to build connections between the school-based program and work, a work-based learning component can complement a student's classroom experience.

STWO and TPAD programs can work together effectively. For example, in Seattle, Washington, representatives from The Boeing Company helped area school districts and community colleges to establish curricula and standards for a manufacturing TPAD program and now operate a work-based internship program for students enrolled in the TPAD program. To be selected as an intern, a student must make a commitment to follow a four-year sequence of applied study, combining two years of high school and two years of community college, culminating with an associate degree. This program demonstrates how a Tech Prep program can successfully combine academics with work-based learning to bring about significantly improved learning and teaching utilizing the principles of applied learning.

Because early reports indicate that education officials are becoming confused about the ways in which youth apprenticeship, school-to-work transition, and TPAD programs all fit together, it is critical that community college officials participate in their state discussions on the development and implementation of a statewide school-to-work system. Special efforts must be made at the federal, state, and local levels to unite these programs and ensure that they work together, rather than against or in competition with each other.

Recommendations

15. Work-based learning should be carefully considered during the curriculum development or revision phase. When determined to be feasible, work-based learning opportunities should be tied to a carefully structured school-based Tech Prep course of study.

16. The TPAD program should be viewed as an essential part of any new school-to-work transition program. Community college officials should join with local and state government officials, business community representatives, and high school officials to assure that these programs work together, rather than in competition with each other.

Marketing TPAD Programs

Important to the success of any given TPAD program is the development of a comprehensive marketing plan, which should promote

TPAD to all program constituencies, including employers, parents, students, and local and state government officials. Marketing materials should emphasize the Tech Prep Associate Degree as both a viable employment credential and a solid foundation for life-long learning. It should also stress the quality of skills, career readiness, and earning potential a student will possess upon successful completion of a degree program. The associate degree in the Tech Prep program will assure students, parents, and employers alike that TPAD is a program of excellence and should be a selling point in all marketing materials.

Recommendation
17. A comprehensive TPAD program marketing strategy should be designed that targets students, parents, employers, local and state government officials, and the general public.

OVERCOMING ADMINISTRATIVE CHALLENGES

Developing, implementing, and maintaining effective TPAD programs require substantial institutional commitment, and federal, state, and local policies to support adequate resources over a sustained period of time. In some states where enrollments at community colleges are burgeoning and institutions' funding formulas are supported by quantity of students rather than quality of education, the move to TPAD programs may be difficult, yet every effort should be made to assure its success. To do so will require the support and commitment of the community college president and governing board, area high school officials, the local employer community, and local and state government officials.

Providing Presidential Leadership

The college president needs to be committed to the goals of Tech Prep and should include appropriate administrators and faculty in the TPAD planning and implementation process. Most importantly, the president should maintain ongoing personal involvement to ensure program stability, direction, and success.

Recommendation
18. Community college presidents should provide the leadership for the development of TPAD programs on their campuses and their feeder high schools. They should be highly involved in educating

the college community, in marketing, in making the necessary connections with area high schools and businesses, and in assigning adequate resources.

Providing Governing Board Leadership

Governing boards, on behalf of their communities and states, should set board policies on mission, purpose, and operation that provide support for effective TPAD programs. They can do so by working directly with local and state boards of education, local and state government officials, business community leadership, and the public at large to develop support for TPAD programs. Boards should periodically review the development and effectiveness of TPAD programs at their meetings.

Recommendation

19. Community college governing boards should take a lead role on behalf of their states and communities in advocating for the establishment and effective delivery of TPAD programs. Boards should provide an appropriate policy framework in support of TPAD program and presidential leadership, and undertake efforts to ensure that adequate support and resources are provided.

A successful TPAD program requires commitment, effort, and willpower from a great many people, as well as substantial resources to sustain it over time. It is a complex undertaking, but one with great potential for rewards, particularly for the nation's workforce of tomorrow. To assist community colleges through this formidable process, papers written by some of the nation's leading Tech Prep experts follow this section. They elaborate on the challenges and rewards community colleges can expect when implementing a TPAD program.

Part V

TECH PREP WHITE PAPERS

GERALD C. HAYWARD

Tech Prep: So Much Promise, So Much Work

S THE READER WILL SOON DISCERN, I AM A STRONG BELIEVER IN the concepts undergirding Tech Prep. I have labored too long in the school reform vineyard, however, not to know that education reform is an extraordinarily complex, mysterious enterprise that has never been and never will be easy. Tech Prep is, in my view, the most promising of the school-to-work initiatives and, if properly implemented, it has the potential not only to improve both the way we prepare young people for work and the way high schools and community colleges deliver instruction to all students. However, the very strength of this concept—its comprehensiveness—often leads to difficulties in implementation. The purpose of this paper is to point out those critical implementation issues that frequently have meant the difference between successful programs and those that have faded from our memories, not because the ideas were wrong but because they were not appropriately and broadly enough placed into practice.

THE EDUCATIONAL AND ECONOMIC CONTEXT FOR TECH PREP

Education Reform

The publication of *A Nation at Risk* by the National Commission for Excellence in Education (Gardner et al., 1983) provided the focal point for much of the school reform efforts of the 1980s. The commission noted that "More and more young people emerge from high school ready neither for college nor for work." Although the commission also said, "this predicament becomes more acute as the knowledge base continues its rapid expansion, the number of traditional jobs shrinks and new jobs demand greater sophistication and preparation" (Gardner et al., 1983), its recommendations focused on improving the quality of that portion of the curriculum normally associated with preparing young people for college. In spite of its lofty rhetoric, the commission's suggested approaches to the problem paid scant attention to non-college bound youth as a target audience or vocational education as a viable part of a school reform strategy.

Vocational education was not seen as part of the solution and, not surprisingly, school districts engaged in very little vocational education reform. This was due to the lack of attention paid to the school-to-work transition in national reform efforts, the notion in some quarters that strong academic preparation was the best preparation for work, and the generally low esteem in which many vocational education programs were held. Throughout the 1980s, school districts experienced significant declines in vocational education programs, due to some extent to increased graduation requirements, which pushed elective vocational education courses from the curriculum. It was in this context that Dale Parnell published *The Neglected Majority.*

Parnell proposed a new program designed both pedagogically and substantively to attract and appeal to the student for whom a baccalaureate degree was not a viable alternative. Parnell called the first iteration of his new program "Careers Education," the delivery system that helps students develop the competencies required to function in the real-life roles of "learner, wage earner, citizen, consumer, family member, leisure-time pursuer, and individual" (Parnell, 1985, p. 65).

Parnell's answer for all this was contained in his comprehensive plan for elementary and secondary schools and community colleges. First, all of basic education would be infused with practical examples for the world of work and life roles. Even students in the primary grades would be able to see the relationship between what they were learning and the usefulness of that learning. Second, career exploration in the middle grades would involve a rigorous, multi-disciplinary approach. Third, adolescents would engage in a new prevocational program that would explore all of the clusters and families of occupations. All students would explore all clusters. The next step would occur in the first two years of high school and would include the development of a common core of learning, including communication skills, social sciences, mathematics, and physical/biological sciences, with a career education emphasis. In the eleventh grade students would choose one of three paths: a college prep/baccalaureate degree major, a 2 + 2 Tech Prep Associate Degree major, or a vocational cluster major. Watered-down general education courses would be eliminated. The last two years of high school in the 2 + 2 Tech Prep option would be closely linked with community colleges, so that all the course work necessary to prepare students well for higher education would have been completed in high school.

Preparation of a Highly Skilled Workforce

Subsequently, a series of important papers focusing on the broad array of skills required for the workforce of the next century were being published. Such reports as *Workplace Basics: The Essential Skills Employers Want* (Carnevale, Gainer, and Meltzer, 1990), *Workforce 2000* (Johnson and Packer, 1987), and the report of the Secretary's Commission on Achieving Necessary Skills (1991) helped concentrate attention on the activities and programs schools can adopt to improve the quality of the school-to-work transition. At the same time, the reports were often critical of traditional vocational education courses. *Workforce 2000* declared that there is no excuse for vocational programs that 'warehouse' students who perform poorly in academic subjects or for diplomas that register nothing more than years of school attendance (Johnson and Packer, 1987). *What Work Requires of Schools* reported that "most schools have not changed fast enough or moved far enough" in response to changes in work and the pressures on the schools of the past decade (Secretary's Commission on Achieving Necessary Skills, 1991, p. 4).

It was realized in the latter part of the 1980s that the nation had to take seriously the responsibility for preparing all students. The first alarm was sounded in *The Forgotten Half* (William T. Grant Foundation Commission of Work, Family, and Citizenship, 1988), which demonstrated a decline in employment opportunities for America's youth, particularly students with a high school diploma or less. Differences in skills, measured by differences in education, are increasingly associated with wage and employment differences across time. However, it does not necessarily follow that increasing the skills of those increasingly disadvantaged in the labor market will affect their abilities to gain employment. To do that depends also on the demand for more highly skilled labor, independent of changes in the characteristics of the labor supply.

Confluence of Educational Reform and Workforce Preparation

The Tech Prep movement, still in its early stages, was given a strong impetus by the timely confluence of the educational concerns expressed by Parnell, who saw Tech Prep as one part of a comprehensive school reform strategy, and the concern expressed by employers and researchers about the quality of workforce preparation and the projected demand for a high-skilled workforce.

In part, the pursuit of this initiative is an outgrowth of education, government, and business leaders reacting to significant economic, technological, and social changes of the 1980s. These include structural changes in the economy linked to fewer industrial production jobs, more service industry jobs, a demand for trained technicians, and the need to improve the quality of education for all students, not just the college bound.

The Tech Prep program was only a part of the overall congressional strategy for the enhancement and improvement of vocational education. Congress's new legislation, the Carl D. Perkins Vocational and Applied Technology Education Act of 1990, represented the most significant policy shift in the history of federal involvement in vocational–technical education funding.

For the first time in federal vocational legislation, emphasis was placed on academic as well as occupational skills, and directed toward "all segments of the population." These new changes in policy were generated by several concerns. One is a strong concern that American firms are losing their competitive edge in world markets. Although there are multiple explanations for such a perceived decline, there is a strong tendency to place the blame on one factor—a labor force with analytical skills that are insufficiently developed for a high-performance work environment. If we assume that the United States is in economic decline and that the cause can be found in the workforce, improving the cognitive skills of the American worker offers the only hope for the preservation of the United States as a high-skill, high-wage economy.

In response to these concerns, Congress, in enacting the Perkins Act of 1990, set the stage for a three pronged approach to better preparing a high-skilled workforce. Perkins emphasizes the integration of academic and vocational education, articulation between segments of education engaged in workforce preparation, and closer linkages between school and work. These three themes also lie at the heart of the Clinton administration's new school-to-work initiative. In a historic degree of cooperation rarely seen in Washington, D.C., the departments of Education and Labor, the two agencies most responsible for job preparation and training, set aside years of often counterproductive bickering to decide on a new course of action for the federal government. The new initiative emphasizes the same three themes, but with important differences in emphasis on at least three dimensions from the congressional reform. First, the Clinton administration proposes to strengthen the school-to-work connection by requiring paid work experiences for every student. The second major distinction is the emphasis the administration places on including all students in the school-to-work program. No longer is Tech Prep seen as a program solely for those students not eligible for college prep. The administration believes that if Tech Prep is seen as a program for non-college bound youth, it will be viewed as just another vocational education program for students who cannot succeed in a rigorous college preparatory program, and it will fail. The third distinction is that the administration's proposals envision a radically different high school than the one that now exists, in which the entire high school and its curriculum are reformed to provide a high school education for every student, much like Parnell's Careers Education program.

ESSENTIAL FEATURES OF TECH PREP

The high school Tech Prep curriculum is designed to prepare students for advanced technical specialization in the community college. Tech Prep runs parallel to college prep, uses a common foundation of mathematics, science, communications, and social sciences to build advanced-skills, is built on applied academics, and uses a competency-based core curriculum structured around a career cluster of occupations. The high school curriculum places a heavy emphasis on building a strong foundation (both academic and vocational), leaving much of the advanced technical courses for the community colleges. At the completion of the program, the student should have completed all the course work necessary to obtain an associate degree in the community college (thus the common appellation TPAD—Tech Prep Associate Degree).

Ramer (1991), using a Delphi process with eight national experts and local groups of educators from community colleges, secondary districts, and regional occupational centers and programs further detailed Tech Prep (2 + 2) components:

Purpose: The purpose of a program is to eliminate unnecessary duplication of course work, thus offering time-shortened curricula and a smoother transition from one educational level to the next.

Participants: The essential participants are the community college and high school. The programs are developed jointly by administrators and faculty of participating agencies.

Criteria for Development: Programs should be developed in an occupation for which the demand for employment is substantiated and the knowledge and skills are greater than can be met in a high school or community college program alone.

Curriculum: Curriculum requirements should include technical skills, written and oral communications, mathematics, interpersonal skills, science, and job-search skills.

College Credit: Students completing the high school courses may receive college credit for that work as long as the basic skill competency requirements for the equivalent college courses have been satisfied. Students who earn college credit for high school courses with a formal articulation program may apply those credits toward the major.

Length of Program: The length of time needed by a full-time student to complete the curriculum depends on the nature of the program.

Program Award: The student completing a program receives an associate degree or certificate of achievement depending on the nature of the program (Ramer, 1991, pp. 3–4).

Articulation can benefit students, educational institutions, and the community at large. Frequently cited advantages to students include better preparation for skilled work, the elimination of duplicate course work, the opportunity to earn college credit while enrolled in high school, and a more efficient use of student time and money. For high schools, one of the major purported advantages is increased student retention. Colleges can, at least theoretically, expect a reduction in the number of remedial or basic courses. Some researchers argue that articulated programs are an effective means of confronting reduced funding for education by enhancing retention and reducing expenditures at the postsecondary level for remediation.

Gene Bottoms, in *Tech Prep Associate Degree: A Win/Win Experience* (Hull and Parnell, 1991) reports that "recent regional and national studies show that well-designed vocational courses can raise academic achievement levels significantly" (p. 381). Furthermore, Bottoms concluded that a planned program of vocational and academic study (sharing the goals of the advocates of integrating academic and vocational education) has the potential of:

> providing a structured and purposeful high-school experience, raising academic and technical achievement expectations for students, motivating students to pursue more rigorous academic courses, and creating a team of vocational and academic teachers (Hull and Parnell, 1991, pp. 385–386).

The concept of Tech Prep incorporates many other important research hypotheses. In summary form, these principles include:

- *Integrating academic and vocational education.* Strengthening academic preparation by emphasizing contextual learning will lead to improved learning (Grubb et al., 1991).
- *Strengthening the connection between the world of work and the world of school.* Reinforcing school work by providing experiences in the workplace that utilize concepts learned in school will lead to improved academic knowledge and work skills (Stern et al., 1991).
- *Enhancing the connection between curricula in high schools and community colleges.* This has important advantages, most importantly, it makes it possible to increase the level of technical expertise of program completers (Dornsife, 1992; Hull and Parnell, 1991).
- *Emphasizing sequences of courses leading to a degree or a certificate.* Students successfully completing sequences of courses and gaining a postsecondary degree and/or a certificate will possess greater technical skills, have greater flexibility to respond to future changes in the workplace, and have better prospects for lifelong earnings (Hoachlander, 1991).
- *Giving students more meaningful educational opportunities in high school that have the prospect of leading to meaningful, well-paying employment.* This will help reduce the high school dropout rate (Hull and Parnell, 1991).

- *Outcome oriented programs with specific competency based curriculum and with specific goals (job placement, or continuing education).* These programs lend themselves to the current emphasis on accountability (Hoachlander, 1991).

KEY COMPONENTS OF TECH PREP

The Tech Prep model is an emerging, ever-changing concept. Nevertheless, it is possible to identify the key components of Tech Prep programs and to further identify how the concept of Tech Prep incorporates many of the important research hypotheses currently under study.

Articulation

The principal defining characteristic of Tech Prep is curriculum articulation. Tech Prep represents an advance over prior articulation efforts in that the articulation occurs between programs or majors (e.g. health, graphic arts) and is not limited merely to courses (e.g. welding, shorthand). Enhancing the connection between curricula in high school and community colleges has important advantages. Most importantly, by reducing duplication and by making expectations clearer, it makes it possible to increase the level of technical expertise of program completers (Dornsife, 1992; Hull and Parnell, 1991; McKinney et al., 1988).

Integrating Academic and Vocational Education

Earlier efforts at articulation (including most 2 + 2 programs) made little or no attempt to integrate the vocational and academic curriculum at either the high school or community college level. Although many Tech Prep programs, especially at the community college level, are only beginning their work to bridge the gaps between academic and vocational education, it is clearly a priority activity (Bragg, 1992). The relative development of this component of Tech Prep programs is often a good indicator of program sophistication. Strengthening academic preparation and emphasizing contextual learning should lead to improved learning and to better prepared students (Grubb et al., 1991).

Enhancing the Connection Between Work and School

Another characteristic of quality Tech Prep programs is the degree to which programs actively involve the business and labor community. The older, more sophisticated programs tend to have better developed school-to-work linkages. Reinforcing school work by providing relevant work experiences will lead to improved academic knowledge and work skills (Stern et al., 1991).

Emphasizing Core Curriculum and Sequences of Courses

Another distinguishing characteristic of quality Tech Prep programs is the notion that all students should have a "core" or base curriculum that includes academic courses of substance in an orderly sequence of courses that builds on a knowledge base of the preceding work. Students successfully completing sequences of courses and ultimately gaining a postsecondary degree and/or certificate should possess greater technical skills, have greater flexibility to respond to future changes in the workplace and have better prospects for lifelong earnings (Hoachlander, 1991; Grubb et al., 1991).

Outcome Oriented Programs

Programs with specific competency-based curriculum and with specific goals (job placement or continuing higher education) can better can be held accountable for success. A distinguishing characteristic of advanced, sophisticated, high-quality Tech Prep programs is their adoption of a competency-based curriculum with accountability mechanisms. Many Tech Prep programs have not adopted outcome oriented programs, but like integration efforts, attention is being paid to this important component (Hoachlander, 1991).

CURRENT STATUS OF TECH PREP

For many of us concerned about the quality of workforce preparation in the United States, there is no program that offers more promise than Tech Prep. Alone of all the models currently in vogue, it is firmly based on the integrated triad of vocational and academic education, secondary and postsecondary education, and school and work. Although it is growing rapidly and although the overall outlook is positive, there are several concerns that must be addressed by the community college community if the concept of Tech Prep is to succeed. In the next few years, the concept of Tech Prep will be "competing" with other school-to-work programs for administration and congressional favor and related funding, so we must identify and deal with areas of particular concern. Again, the point of this exercise is not to decry the lack of perfection in Tech Prep programs across the land, but to serve as a source of constructive criticism about where Tech Prep programs can improve. There now exist many exemplary Tech Prep programs and many marvelous examples of how Tech Prep can work, but there need to be many more. We now know that each of the key components of this program can be made operational. Unfortunately, as yet, only a handful of programs adequately address all of the key components.

Articulation

Prior to the coinage of *Tech Prep* the phrase 2 + 2 was used to describe high school/postsecondary articulated vocational education programs. They most frequently took one of two forms:

- time-shortened programs, in which the primary result of articulation is to shorten the time it takes to complete a specified curriculum, and
- advanced skills programs, in which the primary result is greater technical expertise.

Concurrent enrollment and advanced placement in community college courses are two frequently employed methods of reducing the time it takes to complete a given sequence of courses. The more sophisticated, and rarer, model of Tech Prep is a skill-enhanced model, which provides a more advanced curriculum in an equivalent time period, as a result of the elimination of duplication of course work. Clearly, if the goal is to increase skills, the skill-enhanced model should be the practice. Unfortunately, the vast majority of all the programs currently in operation merely speed up the time it takes to complete the degree or certificate—not a bad outcome, but insufficient if the real goal is to build a workforce with greater skills.

The term articulation, as it has been applied to high school and community college coordination, has referred to the coordination of *courses* between institutions. An important distinguishing characteristic of Tech Prep programs should be that Tech Prep articulation refers to articulated *curricula* or sequences of courses. Results from recent investigations (Dornsife, 1992) indicate the implication of that distinction is lost on many school officials. Schools tend to apply the phrase *articulated curricula* to all articulation agreements whether they refer to individual courses or a sequence of courses. In addition, many schools and colleges refer to virtually any articulated vocational programs as **Tech Prep** as long as the curriculum is associated with vocational or technical programs areas (e.g., business, health occupations, engineering).

Integrating Academic and Vocational Education

If community colleges have trouble with articulation and have not been inclined to adopt the more rigorous, more effective strategy, they have at least some notion of what needs to be done. The same cannot be said for integrating academic and vocational education. There is little evidence at the postsecondary level for this important component of Tech Prep. The division between the academic and vocational segments of the faculty appear to be even greater in community colleges than in high schools. Furthermore, there is little perceived need to alter course. In many states, it is assumed that community colleges need to make few changes to either their curriculum or their pedagogy. The notion in some quarters is that only high schools need be concerned about integrating the curriculum. Nothing could be further from the truth. Many observers are concerned that community colleges will be unprepared to deal with the growing numbers of high school students who will have graduated from high schools with fully integrated academic and vocational programs only to arrive at a community college where little has been done to improve either the curriculum or the pedagogy. Tech Prep will fail without strong curricular and pedagogical changes at both the secondary and postsecondary levels.

Enhancing the Connection Between Work and School

Although there is much that is positive in the working relationships between community colleges and employers, and the level of activity has increased measurably in the last few years, there still is much that can be improved. States and colleges are very uneven in their level of working relationships with the private sector. In some states the community colleges are seen as an integral part of the state's economic development strategy and work closely with the state leadership in integrating their programs with other state job preparation and training strategies. In other states, the community colleges are on their own and depend solely on local initiatives for their private sector connections. Colleges must improve public-private partnerships if they are to be significant players in the nation's workforce preparation plans.

Emphasizing Core Curriculum and Sequences of Courses

Very few community colleges and high schools have entered into truly collaborative curricular reforms as envisioned by the proponents of Tech Prep, in which high school faculty and community college faculty actually design a new curriculum that maximizes the strengths of both segments. High school and community college faculty should agree together on the components of the core as well as the appropriate sequence of courses necessary to complete a technical specialization. Even taking into account the problem of the costs of released time for these collaborations, there are still very few consortia that have even attempted to develop a truly coordinated core curriculum, with appropriate sequences of courses.

Outcome Oriented Programs

Very few consortia have adopted a competency-based curriculum and an equally disappointing few have adopted sophisticated evaluation and accountability mechanisms. The vast majority of programs have in fact paid little attention to student outcomes and, consequently, few have in place evaluation and accountability mechanisms to enable policy makers to determine whether the programs work or not. It can be argued that it is too early to be concerned with outcomes—that the programs are too young to be held accountable for results. However, "early" is exactly the time to be thinking about what kind of evaluation and accountability tools will be utilized. Most programs have concentrated on short range, program implementation problems, to the detriment of longer range, program-sustaining components, like evaluation. As is all too frequently the case with education reform movements, a program with great potential may flounder because of its inability to display, in ways convincing to policy makers, its effectiveness.

We must begin to take Tech Prep seriously and view it, as Dale Parnell did, as a component of a comprehensive, kindergarten-through-associate-degree strategy designed to reform the ways that all schools work for all children. If we fail to do so, all the good intentions of Tech Prep's advocates and all the good work they have done will be for nothing. Tech Prep will become another in a long line of educational programs that promised so much, but delivered so little.

Gerald C. Hayward is deputy director of the National Center for Research in Vocational Education.

WINIFRED I. WARNAT

Tech Prep Education: A U.S. Innovation Linking High Schools and Community Colleges

SINCE THE MID-1980S, THE INCREASINGLY GLOBAL CONTEXT OF THE American economy has forced a significant change in thinking about preparing youth for the workplace. Because education, employment, and the economy are interdependent and considered essential to the nation's economic and social well-being, the United States is aggressively pursuing systemic change and reform in education and workforce preparation. For the first time in U.S. history, education reform is focusing on the majority of students who do not choose to go on for baccalaureate education immediately after high school. For the first time, all youth are recognized as essential to the nation's economic strength.

President Clinton's efforts to stimulate the nation's economic growth and health center on preparing a technically competent, globally competitive workforce. One of his priorities, the newly enacted School-to-Work Opportunities (STWO) Act, aims to transform the decentralized and diverse vocational-technical education delivery system into a cohesive and comprehensive system of systems. It also is designed to provide an array of innovative education-career paths that are based on linkages between secondary and postsecondary education institutions, combining academic and occupational curricula, instruction, and work. Tech Prep education is one of the key education-career path options in this initiative.

THE VOTEC ENTERPRISE

The vocational-technical education (VOTEC) enterprise in the U.S. is large and diverse, encompassing both secondary and postsecondary education. Secondary VOTEC is offered primarily in comprehensive high schools and area vocational schools. The two-year public community and technical colleges are the primary postsecondary providers. Others include area VOTEC institutes, private proprietary schools and four-year degree granting institutions. Work-based VOTEC is provided by employers primarily through coopera-tive education, internships, and structured work experience. In the 1990 census of persons aged 18 through 34 who reported, 43 percent took at least one VOTEC course offered by a two-year public college; 24 percent by a private provider; 19 percent by a VOTEC institute; 10 percent by a four-year college; and 5 percent by an employer (U.S. Department of Education, 1992).

Each student entering high school usually has a choice of three paths of study: (1) the academic path for the college bound; (2) the VOTEC path for the work bound; and (3) the general path for the undecided, a basically unplanned program of study designed around personal interests. Only the academic path has a national student exam, the Scholastic Aptitude Test, needed for admission to many four-year colleges. The VOTEC path prepares the student for entry-level work or continuing occupationally specific preparation at the postsecondary level. The general path is under increasing attack; its survival threatened, because it prepares students for little other than the high school diploma. Success in all three paths is based on satis-factory completion of courses that accrue the requisite number of credits and culminate in the high school diploma.

The VOTEC student going on to postsecondary occupational preparation can work toward a two-year associate degree, an occu-pational certificate, or continue education toward the baccalaureate degree. Two new strategies are leading to significant reform of both the VOTEC and general paths: (1) the movement toward integrating academic and vocational curricula; and (2) Tech Prep education focusing on applied academics and the linking of secondary and post-secondary occupational curricula.

THE EMPLOYMENT SECTOR

Vocational-technical education in the United States is under increasing pressure to provide occupational preparation that address-es current and anticipated workforce needs. Knowing where major job creation, growth, and decline are likely to occur has significant implications when addressing workforce preparation and the recog-nition of occupational skills and competencies.

In the United States, wage and salary employment in manufactur-ing is projected to continue its decline from 19.1 million jobs in 1990 to 18.5 million in 2005, a loss of 600,000 jobs in the manufacturing sector. Between 1975 and 1990, the share of manufacturing employment fell from 23.9 percent to 17.5 percent and is expected to continue downward to 14 percent by 2005. Even so, productivity in manufacturing is projected to continue to increase (Carey and Frouklen, 1991).

Although the annual employment growth for all industries in the service producing sector is expected to be only half that of the 1975 to 1990 growth rate, almost all of the 23.3 million increase in wage and salary jobs for 1990 to 2005 is projected to occur in the service producing sector of the economy. Nearly half of that growth (11.5 million jobs) is expected in the services division, with its two largest industries, health service and business services accounting for one fourth (6.1 million jobs) of the growth. Sixteen of the 20 fastest growing industries and 12 of the 20 industries adding the most jobs are within the services division. Most of these industries are in business, health, social, legal, engineering, management, and educational services (Silvestri and Lukosiewicz, 1991, pp. 64–94).

The Bureau of Labor Statistics provides an analysis of the 30 fastest growth occupations, the 30 largest numerical growth occupations, the level of education required for both, and the 30 largest declining occupations. Fastest growth occupations account for 22 percent of projected overall growth in employment, and largest growth occupations account for 50 percent of projected overall growth. More than two out of three of the 30 fastest growth occupations and nearly half of the 30 largest growth occupations in 1990 had a majority of workers with education and training beyond high school. More than half of the 30 largest declining occupations are concentrated in manufacturing. Key factors contributing to those declines are reduced defense expenditures, increased imports, and advanced technology resulting in higher levels of production. Two overriding factors affecting change in occupational employment are: (1) shifts in employment among industries, and (2) changes in the occupational structures of industries.

TECH PREP EDUCATION DEFINED

Tech Prep education is one of the most significant innovations in the education reform movement in the United States. It provides an alternative education-career path intended to prepare students for high skilled technical occupations, by formally linking secondary and postsecondary academic and occupational curricula. Tech Prep education is geared towards direct entry into the workforce as a qualified technician or continuation with further education leading to baccalaureate and advanced degrees.

Tech Prep education is designed to be available to all secondary students, regardless of gender or ethnicity, including those in college prep, vocational-technical, or general education paths. It also serves those who are economically disadvantaged, disabled, minority, limited English proficient, or have dropped out of school.

A bona fide Tech Prep education program has the following basic characteristics:

- It links in a seamless path the student's transition from high school to college, usually a two-year community or technical college.
- It is based on a planned sequence of study for a technical field beginning no later than the eleventh year in high school and ending with at least one year of postsecondary education.
- It requires a formal, program specific articulation agreement between the secondary and postsecondary institutions.

MODELS OF TECH PREP

There is more than one approach to Tech Prep education, based primarily on the nature of the secondary-postsecondary linkage arrangement. All approaches, however, entail both vertical and horizontal integration. The following five models represent the most common approaches currently being implemented:

2 + 2	The last two years of high school linked by technical preparatory courses with two years of postsecondary education, usually leading to an associate degree or occupational certificate
4 + 2	Four years of secondary school, including two years of pre-Tech Prep preparation before entering the eleventh grade, and two years of postsecondary occupationally specific education
2 + 2 + 2	The last two years of high school with two years of postsecondary technical education followed by two years of higher education, culminating in a baccalaureate degree
2 + 1	The last two years of high school with one year of postsecondary occupationally specific education, culminating in an occupational certificate

2 + Apprenticeship   The last two years of high school with at least two years of apprenticeship training, leading to a certificate

INFLUENCE OF FEDERAL LEGISLATION

The development of Tech Prep programs throughout the United States has been given major impetus by its inclusion and funding in the Carl D. Perkins Vocational and Applied Technology Education Act of 1990. Tech Prep is addressed in Perkins under Titles II and III. The most widely applied Tech Prep education program is the 2 + 2 approach, stimulated largely by Perkins, which also encourages implementation of the 2 + apprenticeship model.

Title II. Through Perkins each state is awarded a basic state grant with multiple uses of funds allowable under Title II. In this section of Perkins, states are given authority to use federal funds for Tech Prep programs providing the programs have:

1. Size, scope and quality to be effective;
2. Integration of academic and vocational education through a coherent sequence of courses; and
3. Equitable participation for special populations.

Title III. Tech Prep receives major emphasis in Perkins under Title III, where it is referred to as the "Tech Prep Education Act." In this section of Perkins, all states are allocated funds to develop and implement local Tech Prep programs as specified in the Act. The two stated purposes are:

- To provide planning and demonstration grants to consortia of local education agencies and postsecondary educational institutions, for the development and operation of four-year programs designed to provide a Tech Prep program leading to a two-year associate degree or a two-year certificate.
- To provide in a systematic manner, strong, comprehensive links between secondary schools and postsecondary educational institutions (U.S. Congress, 1990).

The Tech Prep Education Act (Title III, Part E) is prescriptive and quite specific as to the type of Tech Prep program that is to be developed. First, the Tech Prep program must provide preparation in at least one of the following fields: engineering technology; applied science; mechanical, industrial or practical art or trade; agriculture; health; or business.

A funded Tech Prep education program must also build student competence through a sequential course of study in math, science, and communication, which may be obtained through applied academics. The integration of academics into the Tech Prep curriculum is a basic assumption.

According to the Act, the Tech Prep education program must have the following seven elements:

- An articulation agreement between the high schools and community colleges involved, as well as the other partners in the Tech Prep consortium;
- A 2 + 2 design with a common core of proficiency in math, science, communication, and technology;
- A specifically developed Tech Prep curriculum appropriate to the needs of consortium partners;
- Joint in-service training of secondary and postsecondary instructors to effectively implement the Tech Prep curriculum;
- Training programs for counselors to recruit students, ensure program completion, and subsequent appropriate employment;
- Equal access for special populations to the full range of Tech Prep programs; and
- Preparatory services such as recruitment, career and personal counseling, and occupational assessment.

Although Perkins encourages Tech Prep programs to connect with employers and labor and with four-year baccalaureate programs, those connections are not required. However, states are to give priority consideration to Tech Prep programs that: offer effective employment placement; transfer to four-year baccalaureate programs; are developed in consultation with business, industry, and labor; and address dropout prevention and re-entry and the needs of special populations.

Since the inception of Perkins, Congress has increased the appropriations of federal funds distributed to the states for Tech Prep under Title III for three consecutive years:

Year 1   (91–92) $63.4 million
Year 2   (92–93)   90.0 million
Year 3   (93–94)   104.0 million

This means that to implement Tech Prep education programs, each state received on the average:

Year 1   (91–92) $1.2 million
Year 2   (92–93)   1.7 million
Year 3   (93–94)   2.0 million

IMPLEMENTATION OF TECH PREP PROGRAMS

The fiscal stimulus of Perkins has brought into being over 1,000 Tech Prep programs involving approximately 100,000 students nationwide. According to a survey conducted by the National Center for Research in Vocational Education (NCRVE), most states are building their programs on related past experience (Bragg, 1992). For example, twenty-one states had prior experience with articulation agreements, at least eleven had operated 2 + 2 type programs, and nine had some Tech Prep experience. At least six states have connected their Tech Prep efforts to another major federal law, the Job Training Partnership Act (JTPA), administered by the U.S. Department of Labor. At this stage of development, the linkage between apprenticeships and Tech Prep is minimal.

States making the greatest progress in Tech Prep implementation, thus far, include Illinois, Maryland, Oklahoma, Oregon, and Texas. Tech Prep programs in these states have in common the following characteristics, to which their success can be largely attributed:

- High enthusiasm for Tech Prep from their localities
- Effective collaboration between secondary and postsecondary consortia members
- An extensive marketing and outreach strategy used to attract students, while it informs instructors and parents about Tech Prep
- Significant time and attention given to developing the Tech Prep program's philosophy and vision
- Curriculum development that is specifically designed for Tech Prep and integrates academic and technical knowledge and skills
- Extensive, ongoing professional development in Tech Prep and teamwork for secondary and postsecondary instructors, counselors, and administrators
- Active involvement of employers
- Critical support from key policy and decision makers at state and local levels (Bragg, 1992)

Three Tech Prep education programs, recognized by the U.S. Department of Education through its "Excellence in Tech Prep Award," illustrate quality Tech Prep programs that work are: Partnership for Academic and Career Education Consortium (PACE), Pendleton, South Carolina; Portland Area Vocational Technical Education Consortium (PAVTEC), Portland, Oregon; and Quad-County Tech Prep Consortium, Fort Pierce, Florida.

The purpose of PACE is to prepare students for careers in midlevel technologies. Beginning in the ninth grade and ending with the completion of a postsecondary certificate, diploma, or associate degree, the program provides students with a seamless pathway between the secondary and postsecondary levels of education. PACE was established in 1987 as a business and education consortium to deal with excessive dropout rates and inadequately prepared graduates. Its partners include seven school districts, Tri-County Technical College, the National Dropout Prevention Center at Clemson University, and local businesses and industries. The PACE Tech Prep program links high school and two-year college programs and provides academic and vocational preparation for midlevel technology careers in industrial/engineering technology, business, health, and public service fields. PACE helps each participating school district in restructuring curricula and designing modules in applied mathematics, English, and science courses. Employer participation in PACE has enhanced the tripartite nature of the consortium by producing employer/instructor networks, mechanisms, and materials.

PAVTEC, a 2 + 2 program, was established in 1986, after three years of planning, by Portland Community College and 12 school districts, including 26 high schools within five counties. This partnership also includes private industry, labor, and other educational organizations. The PAVTEC Tech Prep program prepares students for careers in aviation maintenance technology, software engineering technology, medical technology, and other technical fields. In addition to regular participation in strategic planning and in curriculum development and revision, industry has provided summer internships for vocational-technical teachers, counselors, students, and administrators in PAVTEC. During the school year, students are provided with work-based learning opportunities through cooperative education.

The Quad-County Tech Prep Consortium was established in 1991, in direct response to Perkins, to organize and deliver Tech Prep opportunities to students who rank in the middle 60 percent of their classes and have not yet defined specific career goals. The consortium consists of Indian River Community College and all eight high schools of four surrounding local school districts (counties of Indian River, Martin, Okeechobee, and St. Lucie) as well as over 200 local business and industry partners (represented by the Private Industry Council). They developed a 4 + 2 model that involves Tech Prep curricula in four major cluster areas—agribusiness, allied health, business management, and industrial technology, culminating in an associate degree or technical certificate. Integration of academic and technical skills is central to the Tech Prep curriculum design. Business and industry have been heavily involved in curriculum and staff development. During the first year of implementation, 3,400 students were enrolled in Tech Prep.

CONCERNS, CHALLENGES AND CONSIDERATIONS

Even though the localities are responding to Tech Prep education with great enthusiasm, designing and implementing quality Tech Prep programs requires significant investment of time, attention, and commitment. Tech Prep consortium members must understand fully what Tech Prep education and its critical components are. In particular, consortium should avoid the following pitfalls:

Too narrowly defining the composition of the consortium. "Buy in" into the Tech Prep program begins with the consortium membership. Therefore, it is important that membership reflects all major aspects of the program. In addition to local education agencies and two-year postsecondary institutions, the Tech Prep consortium should include counselors, employers, and baccalaureate granting institutions.

No allowance for pre-Tech Prep preparation. Earlier preparation below the eleventh grade is necessary for many students to qualify for entry into Tech Prep programs. States are encouraged to use basic state grant funds to prepare students in the earlier grades. Many states are investing their own funds for pre-Tech Prep preparation.

Perception of exclusivity. Tech Prep is not a program intended only for academically high performing students; it is also aimed at economically disadvantaged and other special populations. Tech Prep should be available to all who are interested in entering a program and efforts to insure access are encouraged.

Superficial articulation agreements. The Tech Prep program cannot rely only on an articulation agreement that grants credit or otherwise substitutes course work at one level for course work at another. Tech Prep articulation should be based on a coordinated curriculum plan that describes a clear course sequence from at least eleventh grade through certification. The articulation agreement should ensure a seamless bridge between the secondary and postsecondary Tech Prep curriculum.

Inadequately designed Tech Prep curriculum. The Tech Prep curriculum is not merely a conglomeration of courses. It should be carefully and jointly planned, with each course tailored to the specific program of study. It is the epitome of academic and technical curriculum integration.

Nominal involvement of employers. Essential to the success of Tech Prep is its connection to employers. Employers need to be actively involved in the consortium, curriculum development, provision of work experience, and ultimate hiring of the Tech Prep completer. The most successful Tech Prep programs have strong employer involvement.

Lack of involvement of higher education. Four-year and advanced degree institutions also have a role to play. They serve Tech Prep in two ways: (1) in the provision of in-service training for both Tech Prep instructors and counselors, and (2) in building the 2 + 2 + 2 option for Tech Prep students.

As a leading innovation in the preparation of a technically competent, globally competitive workforce, Tech Prep education offers many challenges that address systemic and structural change in schools and colleges. Tech Prep requires breaking out of traditional approaches with new thinking, new alliances, and new designs. Recruitment, retention, and results are important themes undergirding the Tech Prep program design. Within that context, the following challenges need to be addressed by Tech Prep programs:

Public awareness/marketing. Tech Prep is new. Parents, employers, and colleagues need to be informed about it. Students need to be "turned on" by it. Information that sparks interest and support needs to be available.

Professional development. Training is an essential component of a quality Tech Prep program. Both secondary and postsecondary instructors and counselors of Tech Prep need in-service joint training on how it works and how to make it work.

Communication. Tech Prep requires collaboration from all key players. It involves breaking down the traditional barriers between levels of education, between academic and occupational education, between education and employers.

Curriculum development. Tech Prep education demands a new curriculum that highlights integration between secondary and postsecondary education and between academic and vocational-technical education. Tech Prep curriculum is developed through team work.

Pre-Tech Prep preparation. Students need to begin preparation for Tech Prep long before the eleventh grade, when the program begins.

As early as middle and junior high school, the knowledge and skills needed to succeed should be determined and developed.

Employer involvement. Tech Prep cannot work without the sanction, support, and participation of employers. This means that Tech Prep programs need to address existing labor market demands.

Quality/accountability. Tech Prep programs require higher performance standards. Competencies and course credits need to be reconciled. Program evaluation needs to be ongoing.

The ultimate success of Tech Prep will be determined by its results. Primary factors on which results will be based include:

- attainment of a certificate or associate degree;
- successful placement of the Tech Prep program completer in a technical occupation;
- successful transfer to a baccalaureate degree program;
- satisfaction of the program completer, the employer, and the university recruiter with the Tech Prep program.

Tech Prep education is breaking new ground in providing preparation for a quality workforce. Because Tech Prep stimulates new organizational and programmatic arrangements, requiring thinking and strategic planning, it demands a high level of cooperation and collaboration from all partners. Functioning properly, Tech Prep education will provide our youth with a highly desirable pathway that works for their as well as the nation's economic well-being.

The president's School-to-Work Opportunities Act concentrates on systemic innovation and change. It assumes that all the social partners will be "players at the table," including secondary and postsecondary education, public and private sector employers, labor and trade unions, community organizations, and others. These partners must develop a consensus on new collaborative approaches and structures that enhance workforce preparation in the United States. The School-to-Work Opportunities Act highlights the importance of linkage and integration between academic and vocational education, secondary and postsecondary education, school-based learning and work-based learning, education and employers and local, state, and federal agencies.

The School-to-Work Opportunities Initiative is on the fast track. Development grants were awarded to states in fall of 1993, and implementation grants will be awarded in spring of 1994 to a small number of states ready to institute a school-to-work system. It is anticipated that all states will have implementation grants over the next several years. Tech Prep is recognized in the school-to-work initiative as an effective mode for preparing students for high-skill, high-wage jobs. It should be a critical component of any school-to-work system. Some states may even use Tech Prep education as their major school-to-work approach. Clearly, it is still one in which community colleges should be key players.

Winifred I. Warnat is director of Vocational-Technical Education at the U.S. Department of Education.

DALE PARNELL

The Tech Prep Associate Degree Program Revisited

THERE ARE THREE RATHER STARTLING OBSERVATIONS TO BE MADE about the education reform efforts currently going on across the country. First is the lack of attention being given toward improving the teaching process. Second is the neglect of the majority of students who will not likely ever earn a college baccalaureate degree. Third is the sparse participation by community, technical, and junior college personnel in the education reform discussions.

Education reform efforts seem to cluster around a new set of "the three Rs": reporting, or more testing; rivalries, or more competition; and restructuring, or more site-based management. Although many of these reform programs make important contributions, most have not addressed the central issue of improving teaching and learning. The fundamental question can be asked with all of these reforms: When the teacher closes the classroom door, is anything different in the teaching and learning process?

Those who support more reporting focus on accountability through testing, even to the point of developing national tests. They seem to be saying, Let's find out what students know or don't know, and who we can find culpable for their not knowing. Let's develop a set of arbitrary knowledge standards. They worry less about what students can do, just what they know.

Those who encourage rivalries emphasize concepts like "choice" or "voucher" symbolizing a desire to create more competition in education. In 1991, the Carnegie Foundation for the Advancement of Teaching conducted an exhaustive year-long study, and reported its findings in a report titled *School Choice* (1992). According to the report, despite some bright spots like several schools in Minnesota that attracted students by creating innovative programs, no choice program had demonstrated a clear link between choice and improving student learning.

While school choice advocates insist that applying free market competition would make good schools better and force weak schools to improve, the Carnegie report said that there is an unimpressive relationship between school choice and improvement of student learning. On the central issue of student learning, standardized test data fail to demonstrate that students who transfer from the public schools are doing any better in the school of their choice. Clearly, school choice is not the panacea that can miraculously sweep away all difficulties that restrict learning and impede good teaching.

Proposals to restructure education center on site-based management, empowerment of teachers, ungraded primary schools, and longer school years. While there are many salutary aspects to the ideas surrounding restructuring education, much of the contemporary discussion about restructuring seems to concentrate upon a process called site-based decision making. This process assumes that those most closely affected by decisions ought to play a major role in making them. All that is known about total quality management supports this concept. But it is nevertheless distressing that small attention is given to improving teaching and learning in the work of site-based councils.

We must expect that these site-based groups and their important restructuring efforts will give the highest priority to the improvement of teaching and learning. If the restructuring efforts in schools and colleges fail to do so, we can only wonder if it has been worth the effort.

THE NEGLECTED MAJORITY

Closely related to the problem of improving teaching and learning is the fact that nearly all of the individuals writing the "how-to-fix-it" reports are products of the college prep/baccalaureate degree curricula and programs. These individuals seem to see reform strategies only through their own personal educational experiences and react as if all of the population looks and acts like them. They ignore the fact that three out of four students in the U.S. education system are unlikely ever to earn a four-year college baccalaureate degree. Yet, most schools and colleges are operated as though the college prep/baccalaureate degree program is the only definition of excellence in education. By this definition, any other approach to education is viewed as second-rate.

Disturbed by this limiting definition of educational excellence, I devoted most of 1984 to writing *The Neglected Majority*, defining the neglected majority as that group of students who are unlikely to complete a baccalaureate degree program. This book offered a different perspective on education reform and excellence by outlining the Tech Prep Associate Degree (TPAD) program, aimed at preparing "the neglected majority" and other students for the demands of our increasingly complex and shifting economy, and to improve teaching and learning.

This book subsequently initiated what has now become a national discussion and educational policy agenda item. The Tech Prep Associate Degree program received bipartisan support of influential political leaders in Congress, who forcefully advocated that the TPAD concept should become a funded part of the reauthorized federal Perkins Applied Technology and Vocational Education Act.

The Perkins Act of 1990 funded TPAD programs to:

- provide planning and demonstration grants to consortia of high schools and community colleges for development of a four-year (Grades 11-12-13-14) TPAD program leading to an associate degree or certificate
- provide strong and comprehensive curricular links between high schools and community colleges, emphasizing continuity between the last two years of high school and occupationally specific associate degree or certificate programs
- combine knowing with doing in the teaching and learning process, based upon the goal of improving this process, particularly through the integration of academic and vocational education.

Now, some eight years later, successful TPAD programs are beginning to exhibit the following characteristics:

- cooperative collegial partnerships among and between high school and community college personnel, particularly faculties;
- regular involvement of employer and labor representatives in program development discussions;
- higher expectations of students as well as the development of applied academics curricula to help students reach these expectations;
- substance-rich curricular structure that provides opportunities for *all* students to understand the connection between academic and vocational education;

- strategies aimed at changing attitudes about technical education;
- provisions for teacher and counselor in-service staff development programs on TPAD programs;
- development of community/technical college "bridge" programs aimed at preparing adult students who have missed the high school portion of the TPAD program; and
- development of performance indicators that show progress and improvements in student learning.

The states most energetically implementing the Tech Prep Associate Degree program focus unwaveringly on the target, the "neglected majority." The Oregon State Board of Education stated the focus in this way:

It is imperative that education give priority attention to the curricular needs of students who are unlikely to complete a college baccalaureate degree program. We must improve the match between what students need to succeed in work and other life roles and how they are taught. This requires the integration of learning-to-know with learning-to-do, changing how curricula and instructional materials are developed, how content is delivered, and how student learning is assessed.

Early returns from schools that have fully implemented the TPAD program indicate some amazing results. Chopticon High School and Charles County Community College in southern Maryland have demonstrated success of students through the Maryland School Performance Program indicators, and student learning is clear.

Chopticon High School
Maryland Functional Test Scores

	Reading	Math	Citizenship	Writing
1987–1988	93.2	63.5	59.6	67.9
1988–1989	93.1	64.9	70.2	74.8
1989–1990	97.4	79.3	67.1	83.8
1990–1991	97.4	79.8	82.4	55.5
1991–1992	97.0	78.4	82.1	87.7
1992–1993	97.5*	83.3*	85.4*	96.1*

*Highest score in school history

In addition, the student dropout rate at Chopticon has gone from nearly 30 percent over four years to under eight percent in the 1992–93 senior class. Other high schools implementing the TPAD program report similar experiences in reducing the high school dropout rate.

Here is what school superintendent Douglas James of the Richmond County Public Schools in North Carolina has to say about the TPAD program:

> The program has had the greatest impact on secondary education in Richmond county since high school consolidation in 1971...Since the beginning of our Tech Prep program, our average SAT score has increased 46 points, the dropout rate declined dramatically, and the percentage of graduates choosing to attend a community college increased from 24 percent to 46 percent.

Much credit must be given to the Clinton administration and the new School-to-Work Opportunities Act, which is aimed at assisting students in making the transition from school to high-skill, high-wage careers. However, it is disappointing to see the success of the Tech Prep program being nearly ignored in this new national school-to-work emphasis. The TPAD applied academics program and work-based learning programs must be developed together, rather than be viewed as unconnected, separate programs.

Students must be academically and technically ready to enter work-based learning programs. If not, employers may become disillusioned with the process and, as a result, their level of support and commitment could begin to diminish. The TPAD program offers the academic, technical, counseling, and motivational components that are so essential to making the school-to-work program be viewed as a program of excellence.

THE COMMUNITY COLLEGE AND THE TECH PREP PROGRAM

Roughly $40 billion is spent annually by U.S. public and private employers for employee education and training programs, and this figure does not include costs for training in the military, which spends an additional $50 billion per year on education and training. Clearly public and private employers have concluded that more fully developing competencies and related performance of the workforce will determine the future health and productivity of their enterprises.

More and more community colleges are waking to the need to cooperate with high schools in developing curricula appropriate for this new technological world. It is absolutely imperative that community colleges become aggressive in examining, developing, and sustaining quality educational programs to serve that great host of Americans who keep this country working.

One key to this new educational thrust is the associate degree. Although the associate degree has been around for some 85 years, it is not well-known by the general public. The arrival of the new "learning age" and proliferation of technician-level education programs have spurred public interest in this two-year degree.

What is an associate degree worth today? Dramatic statistics released recently by the United States Department of Commerce reveal that the associate degree continues to gain significant ground as an important employment credential. For example, the average monthly earnings difference between the associate degree holder and the high school graduate jumped from $300 to $600 per month difference between the years 1984 and 1990. Over the same period, the difference in monthly earnings between associate degree and vocational certificate holders doubled, jumping from $198 to $400, while the difference in earnings between associate degree and bachelor's degree holders remained fairly constant. Probably the most dramatic earnings difference is between the associate degree holder and the high school dropout; this gap has widened from $773 in 1984 to a whopping $1,200 per month in 1990!

The associate degree program is central to the mission of the community colleges and has obvious economic credibility. Emphasizing the associate degree in a TPAD program indicates to teachers, administrators, students, and society that the TPAD consortium has a vision of what it means to be an educated person and affirms the consortium's commitment to program continuity, coherence, and completion. The associate degree, awarded only for completion of a coherent program of study designed for a specific purpose, also indicates that the holder has developed proficiencies sufficient to prepare for upper-division collegiate work or to enter directly into a specific occupation with confidence. It is the hallmark of the educated worker who will be the backbone of tomorrow's work force, and community college leaders must give clear signals about the quality of this degree.

TPAD's Cost-Effective Approach

The TPAD program is cost effective for the student as well as the taxpayer. Most community college students take longer than two years to complete a two-year program. Some students arrive at the college with such severe academic deficiencies that they must take an additional year or two of study to compensate for their deficiencies. TPAD programs will allow them to prepare more fully, overcome any deficiencies while still in high school, and thereby reduce the extra time required in college.

Taxpayers will not be asked to support, via tax dollars, an extra one or two years of college for a host of students. Since most community college administrators estimate that one third of their enrollment may be found in developmental education, TPAD's preparation of students can redirect the dollars supporting developmental education to other, more efficient uses. Community college leaders should stress the cost savings of the TPAD program.

A NEW DEFINITION OF EXCELLENCE

If we are serious about dealing with individual differences in opendoor high schools and community colleges, we must cultivate goaloriented programs of excellence flexible enough to match students' diversity. But educators cannot hope to develop an appropriate educational response unless educational excellence is redefined. We must reject the idea that excellence can be found only in certain university-oriented programs or in preparing for certain professions.

How can educators meet that great range of individual differences among students—rich or poor, able or disabled, university-bound or destined for community college, apprenticeship, military, or a specific job including homemaking? We must vigorously challenge the assumption that a college baccalaureate degree is the sole road to excellence, respect, and dignity. We cannot allow ourselves to confuse social and educational status with equality of opportunity and individual achievement, regardless of the field of study.

Clearly, American education requires some new definitions of excellence—definitions that will hold meaning for all students. The classical orientation toward a baccalaureate degree must change for the three out of four high school graduates who are not likely to complete a college four-year degree. These students can experience educational excellence, but they need more than high-school "bachelor living," "arts and crafts," and other hobby-type courses, and they certainly need something other than unconnected theory courses.

Some fundamental shifts must be made in school and college programs if the needs of all students are to be met and the universal education enterprise is to be improved. High schools and community, technical, and junior colleges must be concerned with improving the educational program and performance of all students, rather than just some of them, and community college leaders must lead the parade toward excellence.

Everyone wins with the Tech Prep Associate Degree program. Students will develop sound basic skills and knowledge while obtaining an excellent education. They will develop the competence to be able to cope with a fast-changing modern life and do so with confidence. Students will be the big winners.

Employers will gain more highly educated workers than ever before. Skilled worker shortages will be alleviated as the TPAD program becomes widely operational in high schools and community colleges across the country. High schools will have more students stay in school to complete their high-school education, and more students will find satisfaction in their applied academics courses of study. The tone and morale of the high school will improve as more students find themselves engaged in meaningful and substantial education programs.

Community colleges will receive students who are better prepared and spend less time and money on remedial and developmental education programs. Communities and states will win because cooperation at different levels of education will eliminate unnecessary program duplication, provide for greater efficiency, and more fully develop the human resources of each region. Finally, the United States will certainly win by the development of a world-class workforce that will outwork, outproduce, and outsmart the international competition. The greatest resource in our nation, the workforce, will be more fully developed than ever in our history.

Dale Parnell is a professor and director of the Western Center for Community College Education at Oregon State University.

Diana Walter and Anita Turlington

Tech Prep: A Practitioner's Perspective

I N JANUARY 1993, THE PARTNERSHIP FOR ACADEMIC AND CAREER Education (PACE) Consortium in Anderson, Oconee, and Pickens counties of South Carolina was chosen as one of the U.S. Department of Education's nine model Tech Prep sites and awarded a demonstration grant of over $500,000. What does it mean to be a model site? It does not mean we have all the answers or solved all of the problems; only that we are off to a good start and are working hard to make Tech Prep the type of program that our students and our communities deserve.

We still have a great deal of work to do in refining the community college's role as a consortium partner. Postsecondary Tech Prep offers community and technical colleges both important challenges in evaluating their community-oriented missions and real opportunities for change.

To understand the impact Tech Prep will have on community colleges, we must first understand what is actually happening to Tech Prep in high schools. Over the past six years, our Tech Prep initiative has evolved from a classic 2 + 2 approach, emphasizing articulation and applied academics, to a more comprehensive vision for teaching and learning.

Tech Prep in Secondary Schools

The PACE Consortium involves seven diverse school districts comprised of 16 high schools and four regional career/vocational centers. After getting off to a rather slow start in 1987, we quickly realized that implementing Tech Prep would require several things of us. First, program development efforts had to respect the unique needs and goals of each district. Second, organizational structures had to support the development of consortiumwide goals, while allowing decisions about curriculum and other program elements to remain the sole responsibility of site-level administrators. Once we dealt with these two basic requirements, area schools began to feel ownership of Tech Prep.

While each district directs the development of Tech Prep programs for schools within its jurisdiction, there is considerable commonality in the basic approach used for program development. All agree in the underlying concept that Tech Prep is *prep*aration for *tech*nologies— the types of careers that require a high school diploma with vocational training up to and including an applied associate degree, either to enter the workforce or to qualify for advancement. The program is designed to eliminate curriculum gaps and overlaps between the secondary and postsecondary levels of education. Tech Prep parallels the college prep curriculum in high school, providing a strong base of academic and occupational study, and enhanced guidance and opportunities for advanced standing. Thus, students can make a smooth transition from one level of education to the next, and from school to the workplace.

The current program design starts in the ninth grade, although coordination with lower grades in curriculum and counseling is evolving in most districts. After the secondary-school component of the program, students take and complete the requirements for a community college degree and enter the workplace or transfer to a four-year college. Students who choose to stop at the end of the twelfth grade are encouraged to work in a meaningful entry-level position that would allow them to continue their education while working full-time. Students have the option to return to the Tech Prep program full-time.

Our secondary Tech Prep programs emphasize building a stronger academic foundation for all students. Like other programs, our secondary schools eliminated general education courses and replaced them with applied academics, using materials developed by the Center for Occupational Research and Development (CORD) and the Agency for Instructional Technology (AIT). Applied academics have changed pedagogy significantly, largely because of the schools' commitment to professional development. High school teachers have completed semester courses designed to provide basic training in applied methodologies. Beyond introductory training, many teachers have participated in more advanced types of training, in areas such as cooperative learning and integrative learning.

While applied academics play an important role in area Tech Prep programs, many schools encourage students to take traditional college-prep courses as another way of building a strong academic foundation for future study and employment. Since most teachers of applied academics also teach traditional college-prep courses, many report they now use some applied methods in teaching their college-prep courses.

The introduction of applied academics into the secondary curriculum caused many districts to examine and reshape the mathematics and language arts sequences for grades 9 through 12. In curriculum planning sessions, college faculty and consortium staff shared information on entry-level college courses and provided sample materials. Curriculum sequences that emphasize higher expectations and innovative teaching techniques are a valuable outgrowth of Tech Prep program development.

The secondary Tech Prep curriculum is beginning to increase integration between academic and vocational curricula. More collaborative classroom projects involving academic and occupational students, and the inclusion of real-world projects provided by area businesses encourage academic and vocational integration. Faculty exchanges and partnering programs for staff development and in-service activities are producing integrated lesson plans from teams of academic and occupational teachers. One district has recently completed an extensive strategic planning process, resulting in a planned restructuring of the 9th through 12th grade curriculum. The new curriculum will emphasize the integration of academic and vocational education and provide students with opportunities for specialized study through career academies and youth apprenticeship programs that will be linked directly to associate degree programs and baccalaureate options.

Secondary Tech Prep programs are also addressing guidance and counseling. Area schools developed career planning guides containing important career information, planning flow charts, and tips for preparing for postsecondary education and the workplace. Some have similar programs at the middle school level to assess students' career interests, provide better information on midlevel technology careers and Tech Prep, and help them develop four-year plans for high school. As a result of these changes, counselors and classroom teachers are becoming more interested and involved in learning about local career opportunities—particularly in midlevel technology fields. They also are becoming more assertive in trying to obtain better information for their students on these career options.

Because the PACE Consortium emphasizes site-based management within a collaborative framework among the partner sites, a number of program elements have been developed by individual districts that in turn have changed the scope of the consortium's Tech Prep initiative. Among these elements are technical advanced study (TAS) and youth apprenticeship (YA), our area's most sophisticated approach to school-to-work transition activities.

An alternative to early dismissal, TAS provides qualified high school seniors with an opportunity to earn advanced standing within Tech Prep by completing entry-level courses on the college campus. Students pay their own tuition and book costs, and provide their own transportation.

Beginning in 1991, one of the consortium's larger districts incorporated a youth apprenticeship component within Tech Prep to link the last two years of a secondary electronics program with the college's associate degree program in industrial electronics. Developed collaboratively between the school district, career center, area employers, the consortium office, and the college, the program integrates classroom and work-based learning and includes workplace competencies that increase in scope and sophistication throughout the duration of the program. This early work in youth apprenticeship has served as a model for other districts.

THE COMMUNITY COLLEGE ROLE—SURVIVING THE EARLY DAYS

In 1987, Tri-County's president, Don Garrison, formed the PACE Consortium with local school administrators. Its coordinating board continues to provide the overall leadership for Tech Prep in Anderson, Oconee, and Pickens counties. Even though Tri-County administrators initiated discussions about Tech Prep, they never considered themselves in charge. The college's role was to participate actively in working out the vision for Tech Prep, to commit college resources as needed, and to be an active and committed partner in ongoing reform. The college had to counter some public school officials' concerns about college-imposed programmatic changes, as well as some skepticism that Tech Prep was just a clever recruiting scheme. Both sides eventually demonstrated mutual respect for each other and subordinated short-term gains to the needs of students.

Articulation—Building the Foundation for Collaboration
Collaboration meant discussing articulated courses, but the discussions of articulation were not always easy. For example, the col-

lege had to deal with granting college credit for high school courses. Some faculty believed strongly that our image would suffer if we promoted articulation agreements to high school students. They finally realized, however, the college was not granting college credit for high school courses. It was granting college credit for equivalent competencies and skills gained in high school courses, and validating that credit at the college, usually in entry-level courses designed for adult students and some recent non-vocational high school graduates who had not taken occupational courses in high school.

Our next quandary concerned program scope. Should we work individually with each high school and career center, or should we attempt to work out articulation agreements that are consortiumwide? With 16 high schools and four career centers in our region, a school-by-school approach appeared too cumbersome and slow. We invited representatives from area high schools and career centers to participate on articulation committees. They wrote generic agreements that allow any school with the identified high school course equivalents to participate in Tech Prep.

We accounted for diversity in course content and competencies through validation. High school faculty were reluctant to this process at first, but they came to understand that validation was needed to assure students had the necessary skills for success at college or on-the-job. The process is based on teacher recommendation, exams, and transition courses.

Tri-County never tried to dictate to high schools what they should teach, because high school occupational courses are designed primarily to qualify students for work when they graduate, while our entry-level courses are designed to provide the foundation for a post-secondary program. Some variation in the balance between theory and hands-on training was natural, so it was inappropriate for us to attempt to suggest or mandate curriculum changes from local high schools. Thus, even though the focus for articulation was on curriculum gaps and overlaps, not on redefining high school courses, a number of teachers have made changes to address competencies that had not been emphasized in high school courses.

The College as Consortium Resource

Working out the articulation process opened the door for other kinds of curriculum collaboration. At Tri-County, the consortium staff involved college faculty in the dramatic curriculum changes that were occurring in area high schools. College faculty partnered with their high school counterparts and business representatives to develop curriculum materials that taught academic skills in English and math by using local business and industry applications. Faculty on both levels became more aware of what was happening in each others' classrooms, and they became much more aware of the demands of today's workplace. Business representatives gained new understanding of the problems faced by classroom teachers, and they helped teachers understand the skills that local employers are seeking in their mid-level employees.

These activities also led to others. At the college, Tri-County faculty and staff are more willing to be involved in career day activities at high schools and junior highs. More high school applied academics teachers request college tours that explore the relationship between technology and communications skills or mathematics abilities, and other curriculum-related issues. College faculty and staff sit on consortium committees and participate in teacher networks. Materials purchased and housed at the college are available to high school partners. Both high school and college administrators have begun to invite each other and their faculties to attend staff development programs. The college collects and reports data on students who achieve advanced standing. In short, Tri-County has become a true consortium resource. A free give-and-take has developed that allows for an easy flow of ideas between institutions.

While all community colleges do not need to serve as fiscal agents for Tech Prep consortia or house a consortium staff, all can play the critical role of resource institution. Making available faculty, data collection, research, and even publishing capabilities will break down many barriers to real collaboration.

POSTSECONDARY TECH PREP

Because Tech Prep is and should be as much a teaching and learning initiative for postsecondary education as it is for secondary schools, all aspects of the Tech Prep experience on the postsecondary level must be reassessed as well, from curriculum and delivery system through counseling and placement.

Students who emerge from exciting, active classes at the secondary level come to us expecting to be challenged and stimulated. After looking closely at the changes in our partnering high schools, we learned about and began to adopt their very exciting teaching methods, including cooperative learning, integrative learning, curriculum integration, and authentic assessment. In addition, a collegewide curriculum improvement process, based on curriculum integration processes, will address the needs of both students and area employers. Most surprising about this new spirit of experimentation in our region is that it flowed from the high schools to the college—certainly a process that some postsecondary faculty would not have found acceptable in pre-Tech Prep days!

Advising and counseling also had to change. As our high school colleagues integrate career awareness into the curriculum (starting in the eighth grade or earlier) and develop better advising materials, we decided that we had to consider the following changes:

- Keeping faculty and counselors abreast of changes in the workplace by serving on advisory committees with business partners, visiting local industries, to name only some;
- Continuing to develop opportunities for articulation with four-year institutions and promoting them effectively to students; and
- Examining our own advising and counseling procedures, especially for students entering with advanced standing and youth apprenticeship students, to ensure that they continue to be focused on a career goal, not just completion of a degree.

Ultimately, any postsecondary institution that serves as a partner in a Tech Prep consortium will be challenged to absorb Tech Prep into its mission and its strategic planning process. We understand now that all components of Tech Prep on the secondary level are available at the college, meaning we can target all students, not just those who have graduated from a Tech Prep pathway in high school. We view postsecondary Tech Prep as an umbrella concept and a support system for real, purposeful change across our institution.

LOOKING TOWARD THE FUTURE

Whatever our past successes may have been, we see new challenges emerging, particularly in articulation, curriculum development, and staff development.

In articulation, we are challenged to meet the needs of students who arrive at college with advanced standing. We do not use a time-shortened approach, because, at least currently, there are not enough advanced students or enough resources to collapse the curriculum sequences. Instead, under an advanced-skills model, students who enter Tri-County with advanced standing may take additional courses for an advanced technology certificate—"postgraduate" work for an associate degree. Students, including adult students, who complete an advanced technology certificate actually receive an additional credential that employers will easily be able to recognize on a transcript. The advantage for Tech Prep students is that they can earn two postsecondary credentials in approximately the same amount of time it would take to complete an associate degree.

The primary advantages of the advanced technology certificates are they do not discriminate against non-Tech Prep students by treating their entry-level courses as remedial and can be developed relatively easily. They can also be particularly responsive to changing industry needs, and implemented with local board approval. This approach addresses our need for advanced skills in a cost-effective and less traumatic manner than redesigning associate degree programs. Finally, it helps us deal with the fact that the numbers of high school graduates entering with advanced standing are often not high enough in any given curriculum, or consistent enough from one year to the next, to justify and sustain a major overhaul of associate degree programs.

The Tech Prep initiative also resulted in more interest in articulating occupational associate degree programs with senior institutions, and ensuring that technical advanced placement credit awarded to high school students will be recognized and accepted for related baccalaureate majors. College faculty are now working closely with faculty and administrators from a local university to develop a model for such a 2 + 2 + 2 articulation process. In addition, the college is expanding efforts to work with regional senior colleges to improve the transferability of occupational associate degree programs. Fortunately, the interest of senior colleges in developing such agreements has been very positive.

Another challenge facing us and all other community and technical colleges is to modify the successful principles and strategies of the Tech Prep initiative for adult students. Adult Tech Prep is already underway here with two bridge/orientation courses that incorporate applied academics for adult students, but we must also improve the use of applied methods by all faculty and redesign our career counseling techniques for returning adult students. Technical advanced placement (articulation) of Tech Prep must be revised to provide appropriate advanced standing opportunities for adults, as well as opportunities for course exemption through new approaches to assess experiential learning.

The Clinton administration's School-to-Work Opportunities Act of 1993 asks us to work more closely with our high school and business partners in developing programs, particularly youth apprenticeship, that will involve students in meaningful and structured work-based learning experiences. We also foresee the great impact that youth apprenticeship will have on the college. Five of the seven districts in the PACE Consortium are actively pursuing the development of youth apprenticeship programs, all linked with associate degree study. The development of postsecondary workplace competencies, evaluation of work-site learning, ongoing communication with work-site mentors, and development of certificates of mastery are just a few of the ways in which Tri-County will be impacted by youth apprenticeship.

Many community colleges will become involved, as we have done, in teacher training. New high school teachers still have little or no exposure to Tech Prep or the concept of applied methodologies, yet they may be assigned to teach applied courses. Through state grants, we are able to offer a recertification course at the college in applied academics by employing both a Tri-County faculty member as a lead instructor and a master teacher from an area high school. The results are very encouraging for new teachers of applied academics, and the process is strengthening the ties between college and high school faculties.

Outcomes

Because we are in the midst of a consortiumwide external evaluation process as part of our U.S. Department of Education demonstration grant, we have only limited data on which to report program outcomes. Some outcomes and impacts are already clear, and some can be safely predicted. For example, a survey of consortium schools in March 1993 indicated that there were 4,254 Tech Prep students in the 1992–93 school year. We expect that the number will grow in coming years. Such high enrollments will affect the college in a variety of ways. Even now we know that the average age of our students at the college has dropped from 27 to 26 years of age. While it is still too early to attribute the change solely to the influence of Tech Prep, it is a factor. The number of area high school graduates entering Tri-County directly from high school grew 38 percent between 1986 and 1992, in a gradual and steady increase. The fastest growing segment of our student population is recent high school graduates.

We can also see growth in the number of technical students. There was approximately a six percent increase in occupational degree program enrollment from fall 1992 to fall 1993, whereas the enrollment patterns of recent high school graduates showed approximately a 15 percent increase. It seems logical to conclude that the Tech Prep initiative has had an impact on the types of majors that recent high school graduates select upon entering Tri-County Technical College.

The number of high school graduates going directly into postsecondary technical education is increasing. In fact, the percentage of area high school graduates enrolling in any state technical college increased from 18 percent in 1990 to 22 percent in 1991. Both years represent the highest percentage in the South Carolina Technical College System.

The PACE Consortium and the Tech Prep initiative have contributed to two spin-off partnerships between the college and area school districts that will significantly impact the college. One partnership focuses on sharing resources and support for a total quality education initiative in each of the seven districts and the college. The other partnership delivers adult and community education services jointly. Organizational structures for the latter partnership were based on those used in the PACE Consortium.

Perhaps a less tangible, but certainly no less valuable outcome is a new spirit evident at the college. It is an adventurous spirit, accepting of change and willing to experiment. Certainly not everyone at the college level has bought in or even fully understands the implications of Tech Prep, but understanding, acceptance, and excitement are growing, as is the conviction that Tech Prep will lead to needed changes at the college level.

Most of all, there is confidence that the philosophies of the Tech Prep movement—making learning relevant and connected, integrating academic and vocational study, linking the classroom to the workplace, helping students make informed career choices, helping teachers become facilitators of learning and better advisors—are valuable for all students, not just those in high school. The changes and reforms involved in the Tech Prep movement have implications for all community and technical colleges. We must be committed for the long haul and for the right reasons. While developing Tech Prep is not always easy, in the end everyone wins.

Diana Walter is executive director of Partnership for Academic and Career Education, and Anita Turlington is postsecondary Tech Prep coordinator at Tri-County Technical College.

CARVER C. GAYTON

Tech Prep:
A Business Perspective

ALTHOUGH I WRITE THIS PAPER AS A BOEING COMPANY EXECUTIVE, I feel comfortable stating that the challenges Boeing faces as a global industry can be generalized to other large, high-tech manufacturing companies throughout the nation. If The Boeing Company is to remain globally competitive and maintain its position as the number one commercial airplane company in the world and the number one exporter in the United States, the company must look at more and better ways to cut costs and, at the same time, improve our processes, products, and services.

In the past, Boeing has confronted several major cost challenges. Among them, the European government subsidies for Airbus Industries a European commercial airplane manufacturer, in the face of no U.S. government subsidies for Boeing products have made the playing field in the commercial airplane business between Europe and America uneven. While the U.S. government is attempting to resolve this issue with Airbus, Boeing must compete despite obvious cost advantage Airbus has in manufacturing its product.

In addition, the costs related to environmental regulations, including record-keeping, training, and compliance costs have increased 115 percent over the past two years. This point is made not because Boeing is averse to environmental regulations, quite the contrary. This is merely another example of an expense that must be factored into the price of Boeing's airplanes. Other structural costs have also increased; health costs for our employees increased 55 percent between 1988 and 1991, and there are no indications, at least in the near term, that such costs will not rise further.

We face other problems as well. Basic skills of new employees, particularly those coming directly from high school, are less than we need. For example, a comparison of recent math and science test scores of students at the ninth grade level in industrialized nations places America's students near the bottom. The skill level of a significant segment of the nation's future workforce should be improved, and, at the same time, the kinds of needs workers have are changing.

At least 70 percent of new entrants into the workforce between the mid 1980s and the early 21st century will be primarily women, minorities, and immigrants. Two questions emerge from this statistic: (1) Are nontraditional employees prepared to meet the higher-skill standards of many of today's high-tech industries? and (2) Are managers prepared to take advantage of the value that can be added to products and services from the diverse perspectives of these new employees? Both questions provide strong challenges to educational institutions and industry.

Because 70 percent of high school students are going directly to work or community colleges, rather than four-year institutions, and up to 50 percent of those who do enroll in bachelor degree programs may never graduate, more energy must be directed toward such students—Dale Parnell's neglected majority—to enable them to transition more easily into the workplace. Also, because the rate of poverty for children is three times greater in the United States than other major industrialized nations, the U.S. cannot compete globally with countries like Japan and Germany, which generally do not have to expend resources to deal with poor, undereducated, underskilled citizens.

INTERNAL ACTIVITIES WITHIN THE BOEING COMPANY

At The Boeing Company, we are not passively bemoaning the fact that the previously stated problems are upon us. There is a sense of urgency, however, that is permeating the entire organization. If we expect to remain the number one commercial airplane company in the world twenty years from now, we must quickly change the way we operate. To this end, we are promoting programs that focus on:

- working with teams;
- hiring and developing employees who can adapt to rapidly changing technology;
- developing managers as coaches and leaders; and
- operating within a continuous quality improvement environment.

Team Building

From a practical point of view, working in isolated, functional areas at the outset of projects creates an atmosphere of turf building, intrigue, distrust, and ultimately more costly products and services. The new 777 commercial aircraft under development by Boeing is being designed in teams of customers, engineers, finance workers, and technicians. This approach has cut engineering design costs in half compared to similar work for our last commercial airliner, and it has established a model for all operations within the company. At the same time, the new approach has implications for the skills and attributes we seek from our current and future employees. In addition to specialized skills, we now expect employees to communicate well both in writing and orally, possess well-developed interpersonal skills, be aggressive listeners, and have a general understanding of a broad array of disciplines—or, at the very least, be quick learners.

Hiring and Developing Employees Who Can Adapt to Rapidly Changing Technology

Within a high-tech environment like The Boeing Company, technology tends to change faster now than in past decades. Employees must have core competencies, problem solving skills, and higher-level thinking skills to learn how to adjust to new technologies. Employees must be prepared intellectually and emotionally to deal with the fact that they may not have the same work assignment two years after they enter the job, or it probably will be broader in scope.

A good example of changing technology affecting every level of the company is the computer. Virtually every engineer must be knowledgeable of computer-aided design, since every new commercial airplane model is designed primarily by computer; the factory worker must operate and maintain numerically controlled machines; and secretaries must be able to operate some type of computer. Less than ten years ago, the computer was not nearly as pervasive at Boeing as it is today.

Developing Managers as Coaches and Leaders

The autocratic management style advocated by theorist Frederick Taylor and others, which forced employees into narrow, inflexible jobs dictated by managers, has become increasingly counterproductive and outdated. With the impact of technology and the long-overdue Theory Y management style, managers within new industrial cultures have different challenges. Boeing Company managers now bring employees together as teams, gleaning perspectives and knowledge from each employee and enhancing operations. Managers must be aggressive listeners, providing leadership to ensure that team and company goals are realized, and identifying attributes of employees that can empower employees as well as the team.

Operating Within a CQI Environment

The "one best way" environment of the scientific management era has been turned on its head within companies like Boeing. The focus now is on the continuous quality improvement (CQI) perspective that products, services, and processes can always be improved. If we listen to our customers, who want continuously higher quality at less cost, we must always improve. This simple, yet revolutionary philosophy emphasizes improving processes by reducing variation. Managers who concentrate on process move away from identifying peoples' errors to correcting process errors, thereby creating a non-intimidating environment for employees. Additionally, employees are provided with simple tools—statistical methodologies—to correct their own work processes, which in turn empowers and encourages them to do their best work.

These new approaches are beginning to have a significant, positive impact on the entire culture of the company. They also have serious implications for the desired skills and attributes of graduates who will enter companies like Boeing. For these reasons, we want to change the company's role in working with schools and colleges from being just a good corporate citizen to being actively involved in educational processes.

EXTERNAL ACTIVITIES IN WHICH THE BOEING COMPANY IS INVOLVED

Beyond the previously stated approaches, society has an important challenge: correcting the ills of the nation's education system, particularly within the K–12 arena. Because of concerns with the educational system, The Boeing Company initiated a variety of programs involving schools and colleges, of which there are too many to list here. Over a nine-year period, improving education has been the company's number one external priority. Boeing's CEO and chairman, Frank Shrontz, is personally at the forefront of many of these efforts. He said, "There are no quick fixes for the problems that affect our schools. The challenge is simply too big for any one sector of society to tackle alone. Educators cannot do it by themselves; neither can parents and students; nor businesses; nor even the combined resources of our local, state, and federal governments. The solution is partnerships that bring together individual citizens and groups from the public and private sectors."

One program that deals with educational issues through the partnership approach is the Tech Prep Applied Academics program, which was launched by The Boeing Company three years ago. Applied academics classes provide an alternative to students in high school who are focusing on something other than a college preparatory program. Rigorous, competency-based, hands-on, applied academics courses in physics, math, communications, biology, and chemistry prepare students for the manufacturing workforce. Students who take these courses usually do not intend to complete a bachelor's degree. Many will instead seek the associate degree offered at community colleges, where graduates who know how to apply these concepts are valued. This is what we regard as the Tech Prep connection. Through the Tech Prep model, students can take applied academics courses along with vocational-technical classes in their junior and senior years in high school and receive credit toward the completion of their associate degree, without needing additional terms for remedial work or fundamental skills. The Boeing Company is convinced that Tech Prep's integrated vocational and academic studies program is the most comprehensive way to prepare work-ready employees.

Thus far, Boeing has been involved in two phases of Tech Prep. In the initial phase, Boeing helped to build the applied academic foundation in the secondary-school system. During the second phase, we helped to promote the development of a statewide Tech Prep Manufacturing Technology Degree program that was recently approved by the State Board for Community and Technical Colleges, and we continue to provide a work-based, student internship program related to manufacturing technology.

Phase 1—Building the Foundation (1990 to 1993)

First, the company provided over $3 million in funds to various high schools and local community colleges in Washington to establish applied academic programs and develop articulation agreements. Specifically, these funds were used as follows:

- Seed grants were given to 60 high schools throughout the state of Washington to implement applied academic programs in principles of technology (applied physics), applied mathematics, and applied communications.

- Articulation grants were awarded to community colleges for developing Tech Prep curricula in partnership with high schools that would allow high school juniors and seniors to take courses for credit toward an associate of arts degree.
- Boeing developed a summer high school teacher internship program that gives applied academics teachers experience in a manufacturing workplace environment that can be taken back to the classroom.

Phase 2—Developing a Manufacturing Technology Degree Program (1993 to Present)

In December 1992, Boeing and representatives from other industries, as well as labor, education, and state government formed an ad hoc committee to promote and support the development of a manufacturing education program for Washington state's existing and future workforce. This group is assisting community and technical colleges develop a manufacturing technology degree program that will teach students the broad, basic skills required to function effectively in today's increasingly complex and competitive manufacturing organizations.

The group's activities include:
- Identifying basic manufacturing entry-level skills;
- Soliciting involvement of other manufacturing firms in the state;
- Advising secondary schools, and community and technical colleges on a core curriculum that responds to industry's needs;
- Determining methods of measuring students' attainment of competencies; and
- Developing a recommended process by which industry can become involved effectively in Tech Prep.

The manufacturing technology program is designed to begin to develop broad basic manufacturing skills in high school—skills that are applicable to manufacturing and other fields as well. This should allow flexibility in career choices as the students expand their knowledge and begin focusing on their areas of interest. We want to keep the beginning course of study broad and basic, particularly at the high school level, so that a student does not have to choose a narrow career pathway before he/she has the time and opportunity to assess all possibilities. Thus, the applied academic courses and the manufacturing lab training are designed to provide the student with basic skills applicable and transferable to many areas of the real world of work.

In February 1993, The Boeing Company approved a summer internship program for students enrolled in the manufacturing technology program. This program provides students with three progressive summer sessions, beginning after eleventh grade, in which they are introduced to career opportunities in manufacturing, taught basic factory skills, and advised on selecting specialty fields within manufacturing. The sessions are being coordinated with the high schools and colleges to ensure that the instruction complements the students' academic courses. The intern program began in the summer of 1993 with 25 students, and it is expected to reach nearly 300 students by 1997. During the summer of 1994, 100 students will be enrolled in internships. Concurrent with the student intern program, the company will continue the teacher internship program for secondary and two-year colleges that was started during Phase 1. Boeing's investment in the internship programs during the next five years will exceed $3.4 million.

A number of essential things are happening as a result of our efforts:

Close communication is developing between education and the private sector. Programs the education community develop must reflect the kinds of knowledge, skills, and attitudes that employers expect. Employers hire graduates more readily if they have had a hand in shaping what is taught.

There is wide involvement of all players. We are pleased to see high school teachers, two- and four-year college instructors, state vocational staff, union representatives, and private trade schools sitting down together to design these programs.

Academic and vocational faculty discover their mutual needs and concerns. Each department has much to learn from the other and each will find students responding positively as subject matter is reinforced across disciplines. Major walls still in existence need to be broken down!

Equivalent credit is being awarded for equivalent outcomes. I served on a special state legislative task force that recommended certain vocational course work, approved by local districts, be accepted as the academic equivalent to core requirements for entry into the state's university system. The Higher Education Coordinating Board of Washington reviewed and approved this recommendation. As a result, it is expected that students will now enroll more freely in many of the high school vocational courses, knowing that it will not cause the doors to a baccalaureate degree to slam shut.

New course work is being developed, not old ones warmed over.
Tech Prep requires instructors to re-examine totally the content of
what they teach. It may require replacing old content with new, even
adding new courses. It will require community and technical colleges
to review seriously what is taught in the high schools so that students'
precious time is not spent going over ground they have already cov-
ered. Truly competency-based approaches must be developed. If students are sent to a community college with a portfolio of skills
already mastered, it won't be necessary to make those students take
a test to prove what they already know. Community colleges must
also review what and how they teach!

Students benefit from work-based experiences coupled with a
strong applied academic curriculum. Our student internship pilot program in the summer of 1993 demonstrated clearly that a focused on-
the-job experience can open new and relevant worlds to students. One
student's comment expressed the feelings of many, "I learned more
math during my four-week internship than in four years in school."

If Tech Prep is to have a long-range and pervasive impact on schools,
colleges, and industry, considerable work must be done to ensure that
business and industry will be full partners in the process. The Tech Prep
conferences I have attended throughout the nation over the past two
years demonstrate that industry basically is unaware of the Tech Prep
movement, despite considerable enthusiasm within the educational
community. Tech Prep is too important to be the exclusive purview of
educators. Of all the recent educational reform efforts, Tech Prep has
the greatest chance of becoming a truly collaborative partnership
among educators, labor, government, and business and industry.

The Boeing Tech Prep effort provides a model, showing that true
collaboration can work, and all participants can benefit. While our
process is continually evolving, we believe we have made a good
beginning, and we are convinced that we cannot rely on the same old
approaches. We are focusing on real partnerships, real collaboration.

Tech Prep's time has come. The Boeing Company is glad to be in
the position that it can help provide Tech Prep with some well-
deserved visibility.

*Carver C. Gayton is corporate director of College and University
Relations at The Boeing Company.*

RICHARD KAZIS

The Future of Two-Year Colleges in Improving the School-to-Work Transition

T HE ADMINISTRATION-SPONSORED SCHOOL-TO-WORK OPPORTUNI-
ties Act of 1993 attempts to create a comprehensive national
system of career preparation for all young people, not just
those who continue on to a four-year baccalaureate program.
The legislation emphasizes the importance of three sets of linkages or
"integrations": the integration of academic and vocational learning,
the integration of school-based and work-based learning experiences,
and the integration of secondary and postsecondary learning.

There is a presumption in this legislation (as there is in Tech Prep)
that the worker of tomorrow must be a lifelong learner; that this
requires individuals to pursue formal education beyond high school;
and that the first postsecondary step along that way for many who
have traditionally floundered in both the educational and labor markets should be the two-year community or technical college. At Jobs
for the Future, we have seen the same presumption in the scores of
innovative school-and-work programs we have studied. There is a
general belief that the community college, with its vocational bias
and its unique relationship and sensitivity to the local labor market,
will be the most frequent beneficiary of these programs' ability to
motivate and to increase the academic performance of mainstream
high school students.

In many ways, though, this presumption is a leap of faith. Few
programs that are part of the new wave of experimentation with
youth apprenticeship or school-to-work transition or, for that matter, Tech Prep, have reached the stage in their development where
early cohorts have made the transition from secondary to postsecondary programs. Moreover, since these programs begin in high
school and are generally seen by proponents as part of a high-school
reform agenda, by far the greatest attention has been given to get-

ting the high school component right. The connection to postsecondary institutions has been a secondary concern, one frequently left for later.

It is time to probe this almost automatic belief in the good fit between high schools and two-year colleges as partners in the provision of high quality programs linking school and work. Will community colleges want to take on the roles that policy makers and secondary school program developers are designing for them? Will community colleges be eager or reluctant when asked to participate? Will they see school-to-work as a significant market opportunity or as a diversion from their real bread and butter?

A second set of questions must also be asked. Even if two-year colleges want to participate aggressively, will they be able to meet and sustain the high standards of program quality that will be expected of them? Are the community colleges' financial and organizational incentives aligned with—or at odds with—the vision of a seamless, integrated career preparation system for young people beginning in the middle school and high school years and continuing through carefully sequenced, rigorous postsecondary occupational programs and into career employment?

The remainder of this paper addresses these questions by highlighting some of the incentives and disincentives facing community colleges as they evaluate their potential role in the school-to-work transition. I also propose some areas for change within community colleges that might encourage more of them to make a greater commitment to school-to-work transition efforts.

The argument of this paper is based on observations from practitioners with whom Jobs for the Future works closely in the field—in high schools and community colleges, at the local and state levels. I also rely on a set of recent interviews with long-time observers and analysts of the community college.

The argument is necessarily speculative; and given the diversity of financing, mission, strategic positioning, and program emphases among the nation's 1,200 community and technical colleges, it undoubtedly misrepresents the experience of many. However, two general conclusions seem non-controversial. First, there are contradictory pulls and pushes within community colleges that shape their motivation to play a key role in these efforts—and these must be explored more fully if that involvement is to increase significantly. Second, community colleges will have to grapple with some very serious issues related to quality, retention, and finance if they are to play the role currently projected for them in the school-to-work transition.

WHY THE PRESUMPTION IN FAVOR OF TWO-YEAR COLLEGES?

It is no surprise—particularly to this audience—that the two-year college is seen by many as the most appropriate institution for the delivery of technical skill training in this country and, therefore, as the logical postsecondary partner with high schools in new programs linking school and work. As two-year postsecondary institution enrollments have risen dramatically over the past thirty years (from 13 percent of undergraduate enrollments in 1960 to 44 percent in 1991), these institutions have become more occupational in focus. According to a study of 1990 data conducted by MPR Associates, about 60 percent of students in community colleges (and virtually all technical college students) now describe themselves as occupational students (Grubb et al., 1992).

In addition to their occupational focus, community colleges are incredibly flexible and entrepreneurial institutions. They are constantly creating new programs to meet the changing demands of students, local firms, and new public funding streams. In recent years, much of this entrepreneurial activity has been concentrated in what has become perhaps the fastest-growing area of community college activity: the non-degree, customized training component.

A recent survey of community college training activities conducted by the League for Innovation in the Community College found that 96 percent of all respondents (representing over 70 percent of all two-year colleges) provide some workforce training for employees of business, industry, labor and government, the majority of which is customized to meet the needs of local employers (Doucette, 1993). According to Stuart Rosenfeld of Regional Technology Strategies, Inc., "Growing numbers of these postsecondary institutions are emerging as major forces in industrial competitiveness—the first level of support for and channel through which the public sector responds to firms' needs for modernization" (Rosenfeld, 1993, p. 24). While the emphasis on training is uneven among the nation's community colleges, there is little question that for small and midsized firms in manufacturing and many other industries, the local two-year college is now one of the key sources of workforce training.

Given these trends, the institution of the community college appears well-positioned and in many ways well-suited to the task of taking high school students and providing them with a rigorous, sequenced technical education. This is exactly what is envisioned by those who seek to integrate secondary and postsecondary learning into a smooth career pathway—whether they are proponents of Tech Prep or of a broader school-to-work transition model. However, a closer look at both the incentives and disincentives facing these institutions reveals a more complicated and contradictory set of interests.

INCENTIVES ENCOURAGING COMMUNITY COLLEGE PARTICIPATION

It can be argued, as many do, that community colleges have a very real and significant interest in working more closely with high schools to improve career preparation for young people. Proponents of Tech Prep have been emphasizing these institutional motives for several years.

Foremost among these institutional motivations is the prospect of attracting a steady stream of better-prepared students for the full-time community college curriculum. Community college leaders often lament the amount of remedial education they must provide incoming students and argue that they are stuck doing the work that high schools should have already done. They are not exaggerating: the Chancellor's Office in California found that, in the 1989-90 school year, 52 percent of all community college students were assessed at the precollegiate level for reading, writing, and computational skills (Commission on Innovation, 1993). College leaders argue that a more educationally prepared young person could move right into more rigorous technical and other degree programs, enabling schools both to reduce remedial education efforts and to move students into and through degree programs more effectively and efficiently.

Being part of a highly touted national school-to-work initiative makes it easier for colleges to attract bright students who previously would have turned elsewhere for postsecondary learning. In addition to attracting a *different* group from the existing pool of high school students, community colleges would also like to see an *expansion* of the pool of high school graduates who can perform well in occupationally focused programs. If high schools did a better job of preparing *more* students for rigorous postsecondary learning, community colleges would reap the benefit in more and steadier enrollments of younger, full-time students with fewer educational disadvantages. According to the president of one community college system, the introduction of quality programs in the high school years that use work and more experiential learning strategies to motivate performance will create "a new, quality market" for community colleges.

Community college leaders emphasize several other incentives for becoming aggressive partners with high schools to improve the school-to-work transition through new work-based learning efforts. One is driven by the culture of community colleges that emphasizes service, responsiveness, and collaborations with other institutions in the community. Other incentives are more hard-headed:

- In a publicly sponsored and funded system, where two-year colleges compete with other state colleges and universities for public resources, initiatives that move technical and community colleges to the forefront of educational reform—and that promise economic impacts as well—can be of significant benefit in relations with state legislatures and other influential institutions.
- Active involvement in school-to-work transition programs that engage employers in intensive ways can have spill-over benefits to the noncredit and nondegree divisions of the two-year college. They can serve as a marketing wedge. School-to-work can provide entree to new employers—and can help cement the college's position as those employers' educational institution of choice. The spill-over can work in the other direction as well. Employers who begin a relationship with two-year colleges for specific customized training may start to think about ways to improve recruitment and training of new entrants to their workforce.

DISINCENTIVES AGAINST COMMUNITY COLLEGE COMMITMENT

While these incentives seem to be solid motivators, there is no question that there are also many compelling disincentives facing community colleges as they consider making a serious institutional commitment to school-to-work transition programs. In fact, at first glance, the disincentives appear more compelling.

I will focus on three: (1) funding mechanisms that favor a high volume of enrollments over sustainable program quality; (2) organizational incentive structures that lead colleges to focus on potential growth areas other than structured school-to-work transition programs; and (3) lack of structure or background in two-year colleges to develop effective work-based learning components for their own students.

Financial Incentives

It is obvious that community colleges live or die by their enroll-ment figures. Most state funding mechanisms reward volume—i.e., high enrollments. Flat reimbursements in many states per full-time equivalent enrollment create an incentive for colleges to keep the costs of delivering credit hours as low as possible, while attracting as many students as possible.

This has several implications. It encourages colleges to maintain as much flexibility in their schedule as possible, so that students can find courses to take at almost any hour of the day. This flexibility enables colleges to meet the needs of very diverse populations, but it militates against programs that require a carefully structured sequence of courses. Flat reimbursements also work against the provision of expensive counseling and academic supports that students might need in order to stay with their program and finish a degree. And they encourage community colleges to keep instructional staff costs low by hiring part-time rather than full-time instructors. The more part-time the staff, the more difficult it is to control quality and to expect any coordination and coherence among course offerings and sequences.

Retention is a serious weakness of most community colleges and funding mechanisms bear part of the blame. Some observers believe that it may even be financially more beneficial for colleges to let peo-ple churn through the system: it is probably cheaper to educate a first-year student who is taking basic and remedial courses than a sec-ond-year student who is taking smaller, more specialized, and often more technology-intensive (and expensive) second-year courses.

If college administrators are thinking primarily about marginal costs and benefits, there is probably little reason to look to expensive, tech-nology-intensive, highly sequenced programs for their core activity. They certainly do not fit well with the largest segment of most commu-nity colleges' business today. Of course, Tech Prep and other funding streams that supplement state grants can change some of these calcula-tions, and community colleges are quite good at pursuing the chance to deliver services supported by categorical funding streams (e.g., JOBS, Adult Basic Education). But these programs are disconnected from each other and are ancillary, not central to most colleges' self-definition.

Organizational Incentives

Just as there are significant financial disincentives to aggressive pursuit of school-to-work transition as a new profit center for two-year colleges, there are aspects of the history, culture and structure of community colleges that also pose obstacles.

Community colleges are often described as archipelagos with little or no connection between the many different islands. The noncredit, nondegree side of the institution usually shares no staff and no strategic planning with the degree-granting divisions. This tends to isolate those faculty and deans who have experience and interest in developing close relationships with local businesses (i.e., the rapidly growing customized training division) from those in degree programs, even the vocational and technical ones. The tendency of community colleges to grow through the creation and expansion of new organizational units, rather than through the addition of new functions to existing units, creates a set of disjunctures that lessens the depth and breadth of institutional ties with the local labor market.

On a less institutional scale, many faculty members are themselves not that interested in expanding school-to-work transition programs. Many are used to teaching older students who are motivated to learn specific topics. They prefer to see themselves more like university professors than high school teachers and do not relish the challenge of motivating young people to perform. If they are part-time instructors, they are probably juggling several different jobs. They want to minimize their preparation and coordination time (additional time spent on program design and coordination is time they are probably not being paid for).

Finally, the bedrock of the community college identity is access. The desire to maintain a flexible course schedule is not driven solely by economics: it is also an article of faith, of community college ideology. Yet, as we have seen, this structural element of community college scheduling is in tension with the desire to promote more and more complex structured sequences of technical and academic courses.

Community Colleges and Work

School-to-work transition programs have as their central innovation the integration of work experiences and classroom experiences into a coordinated holistic learning program. The workplace becomes a learning place and the experiences there are linked to and strengthen those taught in the classroom.

This idea is foreign to most community colleges and most two-year college programs, even the vocational ones. Although most community college students work—to pay tuition since financial aid is difficult to get, and to support themselves and, quite often, their families—two-year colleges have generally done little to grapple with ways that degree programs and students' work experiences can be better linked.

Enrollment in co-op education programs at community colleges is limited. According to an annual survey conducted by the Northeastern University Cooperative Education Research Center, fewer than three percent of community college students participated in co-op in the 1989–90 school year. (U.S. General Accounting Office, 1991, p. 9). A number of technical fields, particularly in health, require unpaid internships as part of their program. This puts some students in the difficult position of having to work two jobs and attend classes if they are to meet course requirements and pay their bills.

Most of the fledgling youth apprenticeship programs in this country that are experimenting with ways to link school and work-based learning more directly have made progress at the high school level, but have found the two-year institutions into which students graduate to resist such innovations. The colleges are not changing to incorporate (and give credit for) work-based learning in the postsecondary years. The institutions do not see the need—and at the scale of current efforts do not feel any pressure—to change their programs or pedagogies. And, not inaccurately for now, they see these programs as high school reform efforts with few pedagogical implications for two-year colleges.

RECOMMENDATIONS FOR CHANGE

In the end, some of the above incentives and disincentives must be changed if community colleges are to assume a central role in whatever school-to-work transition system emerges in this country. These same dynamics also affect the likelihood that colleges will fully embrace Tech Prep, independent of any other efforts to improve career-oriented links with high schools and their students. I believe that two-year colleges should assume that role, that they are the educational institution with the greatest potential to drive significant reform. But I believe the following changes will probably have to come first—and that these will require noticeable shifts in funding mechanisms, institutional culture, and leadership priorities. Community and technical colleges will have to:

- Increase the commitment to retention and quality;
- Take on the challenge of improving the linkage between work and school for their own students;

- Break down the walls between the various internal organizational units; and
- Turn the school-to-work transition movement (Tech Prep included) on its head and make it college-driven, not high school-driven.

Not surprisingly, given the rich diversity among two-year institutions and their state systems, there are good examples of schools and systems taking on each of these challenges, a few of which are noted in these final pages.

Increasing the Commitment to Retention and Quality

In the past five years, most community colleges have experienced significant cuts in public funding, making their concern about per-pupil program costs even more pronounced. If colleges are to embrace school-to-work efforts in a sustainable way—one that includes sufficient resources for counseling and support services and for staffing with full-time instructors committed to and knowledgeable about the program as a whole—then stable, predictable, increased funding will have to be forthcoming. Without such investment, significant reorientation of college priorities is unlikely.

However, there are two-year colleges around the country—many point to LaGuardia in New York as a prime example—where something different and distinctive appears to be going on. In these schools, one gets a sense of an institutional commitment to making the college into a learning community, with greater programmatic coherence and emphasis on sequences of courses that students must follow in order to earn their credentials. Further research on how these schools deal with the trade-off between flexibility/access and structure/quality may yield important, generalizable lessons.

Bringing Work-Based Learning into the Two-Year College

In a recent study of educational institutions providing training for the sub-baccalaureate labor markets in four very different communities, Norton Grubb and three Berkeley colleagues found very important benefits to a strong co-op education program at the two-year college level. In three of the four communities, co-op was weak or nonexistent. One city, though, had a long history of technical college co-op programs that enable students to alternate semesters at work and at school. In this community, unlike the other three, the researchers found the relationship between educational providers and local employers to be positive and interactive. Feedback loops between employers, students, and instructors worked: students could easily learn who were the "good" employers, and firms got a chance to screen students for appropriate matches. According to Grubb:

> The consequence of the co-op programs...is that the distance between education providers and employers, so notable in other regions, has been effectively overcome. Employers spoke knowledgeably and positively about specific educational institutions, routinely hired students from the co-op program, and displayed none of the indifference to the educational system that we found in other areas (Grubb et al., 1992, p. 55).

As this case shows, it is possible for two-year colleges to be much more creative and effective in making work experience—and paid work opportunities—an important part of their program. The benefits are real: to employers, who come to rely on the colleges for an expanding range of workforce-related services; to the schools, which get a much better sense of what local employers need and want; and to students, who are able to earn some money while they are in school at a job that is related to their course of study with an employer who may hire them permanently once they have completed their program.

Increasing the Synergy Between Degree and Non-Degree Divisions

According to Terry O'Banion of the League for Innovation in the Community College, the noncredit component of many colleges is rapidly expanding into a shadow college that threatens to overwhelm the rest of the institution. Some colleges have been doubling continuing education enrollments annually—and profiting from the investment. This dynamic component of the two-year college could continue to grow divorced from, and sometimes at the expense of, traditional degree programs. Or the two could be more effectively linked to complement each other's strengths. As Jim Jacobs of Macomb Community College puts it, "A good school-to-work system must be combined with a good work-to-school system—and the community college can have a key role in both."

Although the conflict between unionized/full-time staff in traditional programs and the nonunion/part-time faculty in training programs is not easily solved, a lot more can be done right now to break down the high walls between college divisions, at least at the level of strategic planning and managerial coordination. The noncredit division is a prime place for innovation and experimentation in instructional approaches, which might then be transferred to traditional programs. Other efficiencies—in marketing and staffing for instance—are also possible.

Shifting from Passive Receiver of High School Products to Driver of the School-to-Work Transition

One state community college official confided recently that college administrators in his state did not see much advantage in the short-term to working with high schools. Their attitude is that school-to-work (and Tech Prep) are primarily high school reform strategies. As such, they may yield a better product for the colleges, but colleges don't feel there is any real incentive to get involved.

I would argue, though, that school-to-work efforts in this country make a grave mistake by being too focused on the high school years and not enough on the postsecondary ones. In the end, serious technical education is probably best left until after high school graduation. Moreover, if employers are to be brought into these efforts in long-term, significant ways, the community college is the institution where that is most likely to happen. And finally, the community college is probably better-positioned than high schools to assume the linking and coordinating roles required to bring employers, secondary schools and postsecondary institutions into effective school-to-work partnerships.

Consider the example of Project Draft, an innovative program established by Macomb Community College in Michigan. Project Draft trains young people and dislocated workers in auto-body design, and also provides upgrade training for currently employed designers. The program was initiated by General Motors in partnership with its design suppliers, local school districts and the college. There are two distinct components: training for students in grades 7 through 12; and work-based training of Macomb Community College co-op students. It reaches two different segments of the community college population—both new high school graduates and working and unemployed adults.

Macomb Community College serves as the educational integrator, working with industry and the schools to translate industry's needs into a comprehensive, sequenced educational program that combines school- and work-based learning. The college has the connections and the credibility from its customized training activities to secure the industry's trust. And it has the experience and ability to help secondary school partners in curriculum design and staff development. Its industry relationships enable the college to create a structured co-op program that both employers and students treat as a step toward placement in a career. And, because it serves a regional market rather than a small, single-school district, the college is able to coordinate among schools and firms across districts and counties in ways that secondary schools cannot.

In this program, Macomb is putting its many institutional advantages to work in a coordinated fashion. The college is organizing the industry-based demand and the educational providers' response; coordinating its own efforts across the credit and noncredit divide; providing students with paid work opportunities; and bringing along the staff, administrators, and parents in participating high schools. The results are impressive: increased program enrollments in the middle and secondary schools and in the community college drafting program; a doubling in the number of firms offering co-ops to college students in the program; a 95 percent placement rate for successful completers; and speedier promotions for program graduates than other employees of firms where they work.

This program, though not automatically replicable in the automobile or any other industry, provides a vision of how community colleges can convincingly—and perhaps to their long-term benefit—establish themselves, as Jim Jacobs recommends, as the "conduit in the system, the institution within the educational community that maintains the linkage between students, businesses and educational institutions on all levels" (Jacobs, 1992, p. 12).

It is a model that might be able to encourage more community colleges to re-evaluate their relatively passive approach to school-to-work transition efforts. John Fitzsimmons, president of the Maine Technical College System, argues that, in the long run, it will be in the two-year colleges' interest to assert themselves as the intermediary and driver of school-to-work efforts. According to Fitzsimmons, "I tell my peers that we might as well get out ahead now, for someone is going to—and then we will end up losing out." Whether such an aggressive strategy might ultimately change the balance of incentives and disincentives discussed above should be a fruitful topic for debate.

Richard Kazis is vice president for Policy and Research at Jobs for the Future.

Part VI

BIBLIOGRAPHY

Bragg, D. *Illinois Tech Prep Planning Strategies.* Springfield: Illinois State Board of Education, 1991.

Bragg, D. *Implementing Tech Prep: A Guide to Planning a Quality Initiative.* Berkeley: National Center for Research in Vocational Education, University of California, 1992.

Bragg, D., and Phelps, A. "Tech Prep: A Definition and Discussion of the Issues." *Wisconsin Vocational Educator,* 1991, *15* (2), 4–6.

Bushnell, D. *Cooperation in Vocational Education.* Washington, D.C.: American Association of Community and Junior Colleges/American Vocational Association, 1978.

Carey, M.L., and Frouklen, J.C. "Industry Output and Job Growth Continues to Slow Into Next Century." *Monthly Labor Review,* 1991, *114* (11), 45–63.

Carnegie Foundation for the Advancement of Teaching. *School Choice.* Princeton, N.J.: Carnegie Foundation for the Advancement of Teaching, 1992.

Carnevale, A.; Gainer, L.; and Meltzer, A. *Workplace Basics: The Essential Skills Employers Want.* ASTD Best Practices Series: Training for a Changing Work Force. First Edition. San Francisco: Jossey-Bass, 1990.

Commission on Innovation. *Choosing the Future: An Action Agenda for Community Colleges.* Sacramento, Calif.: Commission on Innovation, 1993.

Committee for Economic Development. *Children in Need: Investment Strategies for the Educationally Disadvantaged.* New York: Committee for Economic Development, 1987.

Community College of Rhode Island. *2 + 2 Tech Prep Program Report.* Warwick: Community College of Rhode Island, 1990.

Dornsife, C. *Beyond Articulation: The Development of Tech Prep Programs.* Berkeley: National Center for Research in Vocational Education, University of California, 1992.

Doty, C.R. *Developing Occupational Programs.* New Directions for Community Colleges No. 58. San Francisco: Jossey-Bass, 1987.

Doucette, D. *Community College Workforce Training Programs for Employees of Business, Industry, Labor and Government: A Status Report.* Mission Viejo, Calif.: League for Innovation in the Community College, 1993.

Educational Testing Service. *ETS Policy Notes.* 1992, 5 (2).

Evaluation and Training Institute. *Longitudinal Evaluation of 2 + 2 Career-Vocational Education Articulation Projects. First Year Interim Report.* Sacramento: California Community Colleges, 1991. (ED 335 467)

Fadale, L., and Winter, G. "The Realism of Articulated Secondary-Postsecondary Occupational and Technical Programs." *Community College Review,* 1987, 15, 28–33.

Fullan, M. *The New Meaning of Educational Change.* Second Edition. New York: Teachers College Press, 1991.

Fullan, M. *Successful School Improvement: The Implementation Perspective and Beyond.* Philadelphia: Open University Press, 1992.

Gardner, D.P., et al. *A Nation at Risk: The Imperative for Educational Reform.* Report of the National Commission on Excellence in Education. Washington, D.C.: Government Printing Office, 1983.

Grubb, N.; Davis, G.; Lum, J.; Plihal, J.; and Morgaine, C. *The Cunning Hand, the Cultured Mind: Models for Integrating Vocational and Academic Education.* Berkeley: National Center for Research in Vocational Education, University of California, 1991.

Grubb, N.; Dickinson, T.; Giordano, L.; and Kaplan, G. *Betwixt and Between: Education, Skills, and Employment in Sub-Baccalaureate Labor Markets.* Berkeley: National Center for Research in Vocational Education, University of California, 1992.

Hammons, F. "The First Step in Tech Prep Program Evaluation: The Identification of Program Performance Indicators." Unpublished doctoral dissertation, Virginia Polytechnic Institute and State University, 1992.

Hoachlander, E.G. "Guidelines for Developing Systems of Performance Standards and Accountability for Vocational Education." Working paper. Berkeley: National Center for Research in Vocational Education, University of California, 1991.

Hoerner, J.; Clowes, D.; and Impara, J. *Identification and Dissemination of Articulated Tech Prep Practices for At-Risk Students.*

Berkeley: National Center for Research in Vocational Education, University of California, 1992.

Huberman, M., and Miles, M. *Innovation Up Close.* New York: Plenum, 1984.

Hull, D. *Opening Minds, Opening Doors.* Waco, Texas: Center for Occupational Research and Development, 1993.

Hull, D., and Parnell, D. *Tech Prep Associate Degree: A Win/Win Experience.* Waco, Texas: Center for Occupational Research and Development, 1991.

Hull, W. *Comprehensive Model for Planning and Evaluating Secondary Vocational Education Programs in Georgia.* Atlanta: Georgia State Department of Education, 1987. (ED 284 983)

Jacobs, J. "Education and the New Economy." Unpublished position paper. Warren, Mich.: Macomb Community College, 1992.

Johnson, W.G., and Packer, A. *Workforce 2000: Work and Workers for the 21st Century.* Indianapolis: The Hudson Institute, 1987.

Jobs for the Future. *Economic Change and the American Workforce. State Workforce Development for a New Economic Era.* Washington, D.C.: U.S. Department of Labor, Employment and Training Administration, 1991. (ED 336 535)

Key, C. "Building a Transportable Model for Tech Prep Systems Geared for the Twenty-First Century." Unpublished doctoral dissertation, University of Texas at Austin, 1991.

Levin, H. "Improving Productivity through Education and Technology." In G. Burke and R. Rumberger (Eds.), *The Future Impact of Technology on Work and Education.* New York: Falmer Press, 1987, 194–214.

Levy, F., and Murnane, R. "U.S. Earnings Levels and Earnings Inequality: A Review of Recent Trends and Proposed Explanations." *The Journal of Economic Literature,* 1992, *30,* 1333–1381.

Long, J.; Warmbrod, C.; Faddis, C.; and Lerner, M. *Avenues for Articulation: Coordinating Secondary and Postsecondary Programs.* Columbus: National Center for Research in Vocational Education, Ohio State University, 1986.

Mabry, T. "The High School/Community College Connection: An ERIC review." *Community College Review,* 1988, *16* (3), 48–55.

McKinney, F.; Fields, E.; Kurth, P.; and Kelly, F. *Factors Influencing the Success of Secondary/Postsecondary Vocational-Technical Education Articulation Programs.* Columbus: National Center for Research in Vocational Education, Ohio State University, 1988.

National Center for Education Statistics. *Two Years After High School: A Capsule Description of 1980 Seniors.* Washington, D.C.: Government Printing Office, 1984.

National Center for Education Statistics. *Education at a Glance.* Washington, D.C.: U.S. Department of Education, 1992.

National Council for Occupational Education, Inc. *Occupational Program Articulation: A Report of a Study Prepared by the Task Force on Occupational Program Articulation.* Wausau, Wisc.: National Council for Occupational Education, 1989. (ED 321 795)

Ohio State University. *Partners in Progress: A Report by the Task Force on 2 + 2 Tech-Prep Curriculum for Hamilton and Clermont Counties.* Columbus: Center on Education and Training for Employment, Ohio State University, 1990. (ED 321 092)

Parnell, D. *The Neglected Majority.* Washington, D.C.: Community College Press, 1985.

Partnership for Academic and Career Education. *Final Report 1991–92.* Pendleton, S.C.: Tri-County Technical College, 1992.

Ramer, M. *Community College/High School Articulation in California: 2 + 2 Program Definition and Barriers to Implementation.* Sacramento: California Community College Administrators for Occupational Education, 1991.

Richmond County Schools District. *Tech Prep Results, 1986–91.* Hamlet, N.C.: Richmond County School District, 1992.

Rosenfeld, S. "The Metamorphosis of America's Two-Year Colleges." *Economic Development Commentary,* 1993, *17* (1), 24.

Secretary's Commission on Achieving Necessary Skills. *What Work Requires of Schools: A SCANS Report for America 2000.* Washington, D.C.: U.S. Department of Labor, 1991.

Silvestri, G., and Lukosiewicz, J. "Occupational Employment Projections." *Monthly Labor Review,* 1991, *114* (11), 64–94.

Sizer, T. *Horace's Compromise: The Dilemma of the American High School.* Boston: Houghton Mifflin Company, 1985.

Southern Growth Policies Board. *Turning to Technology. A Strategic Plan for the Nineties.* Research Triangle Park, N.C.: Southern Technology Council, 1989. (ED 309 287)

State Center Community College. *2 + 2 + 2 Equals College Credit Now!* Fresno, Calif.: State Center Community College, 1990.

Stern, D.; Hopkins, C.; Stone, J.; and McMillion, M. *Quality of Students' Work Experience and Orientation Toward Work*. Berkeley: National Center for Research in Vocational Education, University of California, 1991.

Task Force on Occupational Program Articulation, National Council for Occupational Education. *Occupational Program Articulation*. Paper developed for the National Council for Occupational Education, 1989.

Tri-County Technical College. *PACE Committee and Board Members*. Pendleton, S.C.: Partnership for Academic and Career Education, 1990.

U.S. Congress. *Carl D. Perkins Vocational and Applied Technology Education Act Amendments of 1990*. Washington, D.C.: U.S. Government Printing Office, 101st Congress, 1990.

U.S. Department of Education. *Vocational Education in the United States, 1969–1990*. Washington, D.C.: U.S. Department of Education, 1992.

U.S. General Accounting Office. *Transition from School to Work: Linking Education and Worksite Training*. Washington, D.C.: U.S. General Accounting Office, 1991.

Warmbrod, C., and Long, J. "College Bound or Bust." *Community, Technical, and Junior College Journal*, 1986, 57 (2), 28–31.

Wentling, et al. *Technology and Preparation Pilot Test: Year 2, School Year 1990–91: Evaluation Report*. Indianapolis: Indiana State Department of Education, 1991. (ED 339 839)

William T. Grant Foundation Commission on Work, Family, and Citizenship. *The Forgotten Half: Pathway to Success for America's Youth and Young Families: Final Report*. Washington, D.C.: William T. Grant Foundation, 1988.